Lecture Notes in Artificial Intelligence   9444

Subseries of Lecture Notes in Computer Science

More information about this series at http://www.springer.com/series/1244

Irina Rish · Georg Langs
Leila Wehbe · Guillermo Cecchi
Kai-min Kevin Chang · Brian Murphy (Eds.)

Machine Learning and Interpretation in Neuroimaging

4th International Workshop, MLINI 2014
Held at NIPS 2014, Montreal, QC, Canada, December 13, 2014
Revised Selected Papers

 Springer

Editors
Irina Rish
IBM T.J. Watson Research Center
Yorktown Heights, NY
USA

Georg Langs
Medical University of Vienna
Vienna
Austria

Leila Wehbe
University of California
Berkeley, CA
USA

Guillermo Cecchi
T.J. Watson Research Center
Yorktown Heights, NY
USA

Kai-min Kevin Chang
Carnegie Mellon University
Pittsburgh, PA
USA

Brian Murphy
Queen's University Belfast
Belfast
UK

ISSN 0302-9743   ISSN 1611-3349   (electronic)
Lecture Notes in Artificial Intelligence
ISBN 978-3-319-45173-2   ISBN 978-3-319-45174-9   (eBook)
DOI 10.1007/978-3-319-45174-9

Library of Congress Control Number: 2016950362

LNCS Sublibrary: SL7 – Artificial Intelligence

Printed on acid-free paper

This Springer imprint is published by Springer Nature
The registered company is Springer International Publishing AG
The registered company address is: Gewerbestrasse 11, 6330 Cham, Switzerland

Preface

Machine Learning, Multivariate Methods and Interpretation: New Avenues in Neuroimaging

Modern multivariate statistical methods developed in the rapidly growing field of machine learning are being increasingly applied to various problems in neuroimaging, from cognitive state detection ("mind reading") to clinical diagnosis and prognosis. Multivariate pattern analysis methods are designed to examine complex relationships between high-dimensional signals, such as brain images, and outcomes of interest, such as the category of a stimulus, a type of a mental state of a subject, or a specific mental disorder. Such approaches are in contrast with the traditional mass-univariate approaches that dominated neuroimaging in the past and treated each individual imaging measurement in isolation.

We believe that machine learning has a prominent role in shaping how questions in neuroscience are framed, and that the machine-learning mindset is now entering modern psychology and behavioral studies. It is also equally important that practical applications in these fields motivate a rapidly evolving line or research in the machine learning community. In parallel, there is an intense interest and several ongoing efforts focused on learning more about brain function in the context of rich naturalistic environments, beyond highly specific paradigms that pinpoint a single function. In this context, many controversies and open questions exist.

This volume is a collection of contributions from the 4th Workshop on Machine Learning and Interpretation in Neuroimaging (MLINI) at the Neural Information Processing Systems (NIPS 2014) conference; moreover, it also includes three papers that received the best paper award at the Third MLINI Workshop in 2013. The first workshop in these series was organized in December 2011. The MLINI workshop series focuses on machine learning approaches in neuroscience, neuroimaging, with a specific extension to behavioral experiments and psychology, and provides a forum that facilitates cross-fertilization across those fields.

The key objective is to pinpoint the most pressing issues and common challenges, and to sketch future directions and open questions in the light of novel methodology. Besides interpretation, and the shift of paradigms, many open questions remain at the intersection of machine learning, neuroimaging, and psychology. These questions include, but are not limited to the following. Can we characterize situations when multivariate predictive analysis (MVPA) and inference methods are better suited for brain imaging analysis than more traditional techniques? How well can functional networks and dynamical models capture the brain activity, and when using network and dynamics information is superior to standard task-based brain activations? How can we move toward more naturalistic stimuli, tasks, and paradigms in neuroimaging and neurosignal analysis? What kind of mental states can be inferred from cheaper and easier to collect data sources (as an alternative to fMRI scanners) such as text, speech, audio, video, EEG, and wearable devices? What type of features should be extracted from such naturalistic input to detect specific mental states and/or mental disorders?

Exploring these and many other related questions remains the source of inspiration for the MLINI workshop series.

December 2014   Irina Rish
                                                Georg Langs
                                                Leila Wehbe
                                             Guillermo Cecchi
                                           Kai-min Kevin Chang
                                               Brian Murphy

Organization

MLINI 2014 was organized during the Neural Information Processing Systems (NIPS) 2014 Conference in Montreal, Canada.

Workshop Chairs

Irina Rish	IBM T.J. Watson Research Center, USA
Georg Langs	Medical University of Vienna, CSAIL, MIT, USA
Leila Wehbe	Machine Learning Department, Carnegie Mellon University, USA
Guillermo Cecchi	IBM T.J. Watson Research Center, USA
Kai-min Kevin Chang	Language Technologies Institute, Carnegie Mellon University, USA
Brian Murphy	Queen's University Belfast, UK

Contents

Networks and Decoding

Multi-Task Learning for Interpretation of Brain Decoding Models

Seyed Mostafa Kia[1,2,3(✉)], Sandro Vega-Pons[2,3], Emanuele Olivetti[2,3], and Paolo Avesani[2,3]

[1] University of Trento, Trento, Italy
seyedmostafa.kia@unitn.it
[2] NeuroInformatics Laboratory (NILab), Bruno Kessler Foundation, Trento, Italy
[3] Centro Interdipartimentale Mente e Cervello (CIMeC),
University of Trento, Trento, Italy

Abstract. Improving the interpretability of multivariate models is of primary interest for many neuroimaging studies. In this study, we present an application of multi-task learning (MTL) to enhance the interpretability of linear classifiers once applied to neuroimaging data. To attain our goal, we propose to divide the data into spatial fractions and define the temporal data of each spatial unit as a task in MTL paradigm. Our result on magnetoencephalography (MEG) data reveals preliminary evidence that, (1) dividing the brain recordings into spatial fractions based on spatial units of data and (2) considering each spatial fraction as a task, are two factors that provide more stability and consequently more interpretability for brain decoding models.

1 Introduction

Cognitive neuroscientists are generally concerned with discovering answer of *where*, *when* and *how* a certain brain activity contributes to a particular cognitive process. Hence, a multivariate analysis of recorded brain activity, e.g., Electroencephalography (EEG), Magnetoencephalography (MEG), or functional Magnetic Resonance Imaging (fMRI), is considered *interpretable* if it can find accurate and stable answer to *where*, *when* and *how* questions. Therefore, improving the interpretability of multivariate analysis is of high interest in the brain imaging literature [24].

Nowadays, mass-univariate hypothesis testing methods are widely employed to make inference on neuroimaing data [11,17,18]. Despite popularity of these univariate methods, they are generally unable to spot complex interactions between different brain areas [7]. Recent studies tried to find multivariate alternatives to univariate hypothesis testing [16,20], however, classification-based approaches are still the most popular tools for multivariate analysis of neuroimaging data [9]. These approaches go under the name of *brain decoding* and generally use linear classifiers to find evidence for stimulus related information in neuroimaging data. The weights of linear classifiers provide quantitative measurements to assess the relation between each dimension of data, i.e., features, and the underlying cognitive task. However, these approaches suffer from lack

© Springer International Publishing AG 2016
I. Rish et al. (Eds.): MLINI 2014, LNAI 9444, pp. 3–11, 2016.
DOI: 10.1007/978-3-319-45174-9_1

of interpretability due to the high dimensionality of data and high correlation between features [3,12,13].

Currently, there are two main directions in neuroimaging literature to improve the interpretability of multivariate linear models. The first concentrates on model selection in order to increase the stability of brain decoding model. This approach suggests taking into account the stability of models in model selection procedure. For example, [22] computed the correlation between weights of models across different cross-validation runs, and utilized it besides accuracy for model selection in joint accuracy-reproducibility space. Analogous approaches have been proposed in [1,4,6,26].

The second approach focuses on the underlying mechanism of regularization to enhance the interpretability of weights of classifier. The main idea is two-fold: (1) customizing the regularization terms to address the ill-posed nature of brain decoding problems (where the number of samples are much less than the number of features); and (2) to incorporate structural or functional prior knowledge into the regularization procedure. Group Lasso [29] and total-variation penalty [25] are tow effective methods in this direction [23,28]. As an example in the neuroimaging context, [9] by modifying the regularization term of logistic regression, proposed a group-wise regularization term for finding sparse and easy to interpret models. Elsewhere, [10] used total-variation penalty to inject a spatial segmentation prior into the sparse model with Lasso penalty. Similar efforts have been made in [3,12,27].

Despite the mentioned efforts, recently [13,14] questioned the interpretability of linear discriminative models, i.e., weights of linear classifiers, due to the contribution of noise to the amplitude of weights. To address this problem, they proposed a procedure to transform discriminative models into equivalent generative models by multiplying linear classifier weights by the covariance matrix of the input features (see 2.2). Their experiments on simulated, EEG, and fMRI data illustrated that, whereas direct interpretation of linear classifier weights may cause misinterpretation of results, their proposed solution effectively solves the problem.

In this study, we approach the problem of interpretability by employing a multi-task learning (MTL) framework in order to improve the stability and as a result the interpretability of brain decoding models. We are willing to stress two key advantages of MTL over single-task learning in brain decoding interpretation: (1) reformulating the brain decoding problem into a multi-task problem, by defining each spatial unit of data as a task, provides more stability for brain decoding models; (2) learning the pattern of activities simultaneously over spatial units increases the performance of decoding compared to the single-task learning where a number of classifiers are trained separately on each spatial unit.

The rest of this paper is organized as follows: in Sect. 2 we introduce multi-task elastic-net and we show how a brain decoding problem can be recast into the MTL paradigm. Then, in Sect. 3, we present our experimental results on an MEG dataset by comparing the performance and the stability of MTL with single-task learning. Finally, Sect. 4 concludes this paper.

2 Methods

2.1 Notation

Let $(X, Y) = \{(\mathbf{x}_1, y_1), \ldots, (\mathbf{x}_n, y_n)\} \in \mathbb{R}^{n \times p} \times \mathbb{N}^n$ be the n samples of neuroimaging data, e.g., MEG data, where each $\mathbf{x}_i$ is a p dimensional vector of spatio-temporal features throughout presentation of stimulus of class y_i. The goal of brain decoding is to find a function Φ such that $Y = \Phi(X)$. In the linear case $Y = XW$ where $W \in \mathbb{R}^p$ represents the weights associated by a linear classifier to every corresponding element of $\mathbf{x}_i$.

2.2 From Classifier Weights to Activation Patterns

Recently, [13] showed the weights of a linear classifier, i.e., W, are not neurophysiologically interpretable. They illustrated that any interpretation based on W can cause wrong conclusions with respect to the spatio-temporal source of signal of interest. As a solution, showing that for every discriminative model there exists an equivalent generative model, they proposed a procedure to transform the weights of linear classifiers to *activation patterns* A:

$$A = \Sigma_X W \Sigma_{\hat{S}}^{-1} \tag{1}$$

where Σ_X and $\Sigma_{\hat{S}}^{-1}$ represent covariance matrix of X and $\hat{S}$, respectively, and $\hat{S}$ is latent factor representing estimated neural sources.

In fact, an activation pattern is the solution of the equivalent generative model that encodes the strength and polarity of the activity of interest in each dimension of data. Therefore, there is a clear physiological interpretation for activation patterns. In the binary classification setting where there is just one latent factor $\hat{Y}$ estimated by the model, the Eq. 1 can be rewritten as:

$$A = \frac{\Sigma_X W}{\sigma_{\hat{Y}}^2} \propto \Sigma_X W \tag{2}$$

2.3 Multi-task Elastic-Net

Multi-task learning (MTL) has recently received particular attention in machine learning and computer vision literature [30]. MTL tries to learn the underlying relation between tasks simultaneously by extracting common information across them. It has been shown that, in some applications, the simultaneous learning procedure of MTL is advantageous over learning each task independently [8]. Furthermore, splitting a single-task problem into a multi-task problem can effectively change the relative size of samples to features for each task. Thus MTL can provide higher stability by reducing the degree-of-freedom of the solution space.

In this study, we first define a *spatial fraction* as a time-series of each spatial unit of neuroimaging data. For example in the case of MEG data, the time-series measured by each MEG sensor is defined as one spatial fraction of data.

Then, we define each spatial fraction as a *task* in the MTL framework. We consider the MTL scenario of having the same outputs and different inputs for each task [2,8]. Thus, a brain decoding problem can be reformulated as $(X, Y) = \{(X^1, Y), \ldots, (X^\tau, Y)\}$; where each pair of (X^i, Y) defines a traditional brain decoding problem (see 2.1) on just one spatial fraction of data, $X^i \in \mathbb{R}^{n \times p^*}$ represents n samples of data at ith spatial fraction, τ represents number of tasks; and $p^* = p/\tau$ is the number of temporal features at each spatial fraction.

Using this new representation of brain decoding, the multi-task elastic-net (MTEN) optimization problem, as an instance of MTL, can be formulated as follows [5,31]:

$$\hat{W}^{MTEN} = \underset{W \in \mathbb{R}^{p^* \times \tau}}{\operatorname{argmin}} \sum_{i=1}^{\tau} \left\| X^i W^i - Y \right\|_F^2 + \rho_1 \left\| W \right\|_1 + \rho_2 \left\| W \right\|_F^2 \qquad (3)$$

where $\|.\|_1$ and $\|.\|_F^2$ are representing the $l1$ and $l2$ penalties respectively, and $W \in \mathbb{R}^{p^* \times \tau}$ is the MTEN weight matrix. The regularization parameters ρ_1 and ρ_2 control sparsity and smoothness over temporal patterns of spatial fractions, respectively.

The MTEN optimization problem can be considered as an extension of single-task regression with elastic-net regularization [32]. A general specification of MTEN is its shared $l1$ and $l2$ penalties among all tasks. Furthermore, in this setting, the number of temporal features of each task (p^*) is reduced by factor of the number of tasks (τ) with respect to that of the original feature space (p). In practice and using common down-sampling techniques even $p^* < n$ is achievable. Therefore, the input data of each task can be a full rank matrix.

To compute the final prediction of the MTL model, we use a simple averaging mechanism among the tasks. We first define a *decoding-related task* (DRT) set D, as a set of tasks which provide decoding performance over a certain threshold ϕ in the training-set. The threshold ϕ can be decided using nested cross-validation or can be fixed based on some heuristics. After finding DRT members, to compute the final prediction for every sample in the test-set, we compute the mean over predictions of classifiers in D.

Furthermore, considering the fact that decoding models with below chance performance are not interpretable under any circumstances, those spatial fractions that are not effective in decoding should be filtered out from the joint activation patterns. Therefore, we merely use the weights of classifiers in D to compute activation patterns of MTEN. The activation patterns associated to unrelated tasks are set to zero when constructing the full spatio-temporal activation pattern A. To compute the activation pattern of each member of DRT set A^{i^*} ($i^* \in D$), we adopt Eq. 2 as follows:

$$A^{i^*} \propto \Sigma_{X^{i^*}} W^{i^*} \qquad (4)$$

3 Experiments

3.1 Material and Experimental Setup

We tested the proposed method on the first 5 subjects of an MEG dataset where visual stimuli consisting of famous faces, unfamiliar faces and scrambled faces are presented to subjects. The original dataset consists of 16 subjects and it is described in [15][1]. This dataset is also used for DecMeg2014 competition[2]. Same as [19], we created a balanced face vs. scramble dataset by drawing at random from the trials of famous and unfamiliar faces in equal number to that scrambled faces. The raw data is high-pass filtered at 1 Hz, down-sampled to 250 Hz, and epoched from 200 ms before the stimulus onset to 800 ms after the stimulus. Thus each trial has 250 time-points for each of the 306 MEG sensors (102 magnetometers and 204 planar gradiometers)[3].

To illustrate the advantage of MTEN in improving the interpretability of brain decoding model, we conduct three different experiments. These three settings help us to examine the impact of division of data into spatial fractions and employing the MTL paradigm separately:

1. We first pool all temporal data of 306 MEG sensors into one vector (i.e., we have 250*306=76500 features for each sample) and then we use the linear regression with elastic-net regularization to solve the brain decoding problem (we refer to this experiment as EN).
2. We divide the data into spatial fractions, then we employ the linear regression with elastic-net regularization to train a model on each spatial fraction separately (we refer to this experiment as STEN).
3. After dividing data into spatial fractions, we use MTEN to train the decoding model (we refer to this experiment as MTEN).

For selecting DRT members in the second and third experiments, the threshold ϕ (see 2.3) is set to $\mu_{perf} + \sigma_{perf}$, where μ_{perf} and σ_{perf} are respectively mean and standard-deviation of performances computed over all spatial fractions (tasks) on the training set. In all settings, the best values for ρ_1 and ρ_2 were decided using nested cross-validation (CV) to ensure unbiased error estimation [21]. In the inner loop of CV, a grid search on $[0, 0.001, 0.01, 0.1, 1, 10, 50, 100]$ is used to find optimal values for both ρ_1 and ρ_2. MALSAR [31] toolbox is used for training the models. The MATLAB code for all experiments is available at https://github.com/smkia/MTL_Interpretation.

---

[1] The full dataset is publicly available at ftp://ftp.mrc-cbu.cam.ac.uk/personal/rik.henson/wakemandg_hensonrn/.

[2] The competition data are available at http://www.kaggle.com/c/decoding-the-human-brain.

[3] The preprocessing scripts in python and MATLAB are available at: https://github.com/FBK-NILab/DecMeg2014.

3.2   Results and Discussions

Figure 1 compares the performance and the stability of EN, STEN, and MTEN experiments. The performance of classifiers is measured based on the area under the ROC curve (AUC). The stability is quantified by computing the pair-wise correlation between weight matrices across 10 folds of CV (see [22]). The bars and the error-bars are showing the mean and the standard deviation of AUC and correlations over 10 folds of CV.

The annotations below each group of bars are showing the result of two-sample t-test between each pair of benchmarked methods, where $-$, $*$, and $**$ are representing *not significant, significant* with $p-value < 0.05$, and *significant* with $p-value < 0.001$, respectively. All the results are corrected for multiple-comparison using Bonferroni correction. Excluding the second subject which shows completely different behaviour, Fig. 1 highlights the following points:

1. While MTEN and EN have more or less the same performance, MTEN provides significantly better stability than EN.
2. STEN and MTEN provide more stability than EN, supporting the idea that dividing the data into spatial fractions improves the stability of models by reducing the degree of freedom of solution space.
3. Despite their similar stability, MTEN provides better performance than STEN illustrating the advantage of learning all tasks simultaneously in MTL framework.

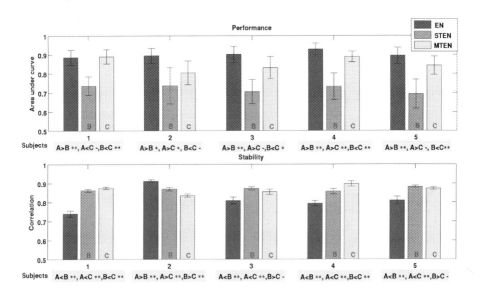

Fig. 1. Comparison between performance (upper diagram) and stability (lower diagram) of EN, STEN, and MTEN for 5 subjects.

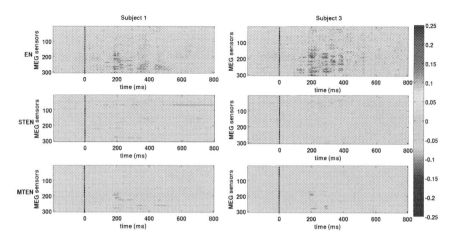

Fig. 2. Spatio-temporal activation patterns of 2 subjects computed by EN, STEN, and MTEN.

Figure 2 elaborates more the advantage of MTL paradigm in improving the interpretability of results. This figure shows mean activation patterns of MTEN, STEN, and EN over 10 folds of CV for two subjects (other subjects show similar behaviour). These activation patterns are computed using Eq. 2 in EN case, and using Eq. 4 in STEN and MTEN cases. The horizontal and vertical axes represent time and sensors dimensions respectively, and the dashed line shows the stimulus onset. Comparison between these activation patterns suggests:

1. MTEN and STEN provide more sparse solution than EN.
2. Activation patterns of MTEN show no stimulus related activity before stimulus onset, in contrast to EN. Considering the experiment design used for data acquisition (see 3.1), any discriminating activity before stimulus onset has no scientific interpretation. These activations before stimulus in EN case can be consequence of overfitting of the model to noise.
3. Pre-stimulus activation in EN case rises the question that the transformation proposed by [13] might not guarantee the interpretability of brain decoding models, and the regularization strategy beside learning algorithm are still playing important roles.

4   Conclusion

In this paper, we introduced a new application of MTL to enhance the interpretability of brain decoding models. Our results on an MEG dataset show that recasting the brain decoding problem into the MTL framework is an effective technique to achieve more stable and consequently more interpretable models. These characteristics of the proposed method makes it more appropriate for making inference in cognitive neuroscience studies. Replacing elastic-net with a new penalization method in the MTL paradigm can be considered a possible future extention for our work.

References

1. Afshin-Pour, B., Soltanian-Zadeh, H., Hossein-Zadeh, G.A., Grady, C.L., Strother, S.C.: A mutual information-based metric for evaluation of fMRI data-processing approaches. Hum. Brain Mapp. **32**(5), 699–715 (2011)
2. Ben-David, S., Gehrke, J., Schuller, R.: A theoretical framework for learning from a pool of disparate data sources. In: International Conference on Knowledge Discovery and Data Mining, pp. 443–449. ACM (2002)
3. de Brecht, M., Yamagishi, N.: Combining sparseness and smoothness improves classification accuracy and interpretability. NeuroImage **60**(2), 1550–1561 (2012)
4. Carroll, M.K., Cecchi, G.A., Rish, I., Garg, R., Rao, A.R.: Prediction and interpretation of distributed neural activity with sparse models. NeuroImage **44**(1), 112–122 (2009)
5. Chen, X., Kim, S., Lin, Q., Carbonell, J.G., Xing, E.P.: Graph-structured multi-task regression and an efficient optimization method for general fused lasso. arXiv preprint arXiv:1005.3579 (2010)
6. Conroy, B.R., Walz, J.M., Sajda, P.: Fast bootstrapping and permutation testing for assessing reproducibility and interpretability of multivariate fMRI decoding models. PLoS ONE **8**(11), e79271 (2013)
7. Cox, D.D., Savoy, R.L.: Functional magnetic resonance imaging (fMRI) brain reading: detecting and classifying distributed patterns of fMRI activity in human visual cortex. Neuroimage **19**(2), 261–270 (2003)
8. Evgeniou, T., Pontil, M.: Regularized multi-task learning. In: Proceedings of the Tenth ACM SIGKDD International Conference on Knowledge Discovery and Data Mining, pp. 109–117. ACM (2004)
9. van Gerven, M., Hesse, C., Jensen, O., Heskes, T.: Interpreting single trial data using groupwise regularisation. NeuroImage **46**(3), 665–676 (2009)
10. Gramfort, A., Thirion, B., Varoquaux, G.: Identifying predictive regions from fMRI with TV-L1 prior. In: 2013 International Workshop on Pattern Recognition in Neuroimaging (PRNI), pp. 17–20. IEEE (2013)
11. Groppe, D.M., Urbach, T.P., Kutas, M.: Mass univariate analysis of event-related brain potentials/fields I: a critical tutorial review. Psychophysiology **48**(12), 1711–1725 (2011)
12. Grosenick, L., Klingenberg, B., Katovich, K., Knutson, B., Taylor, J.E.: Interpretable whole-brain prediction analysis with graphnet. NeuroImage **72**, 304–321 (2013)
13. Haufe, S., Meinecke, F., Görgen, K., Dähne, S., Haynes, J.D., Blankertz, B., Bießmann, F.: On the interpretation of weight vectors of linear models in multivariate neuroimaging. NeuroImage **87**, 96–110 (2013)
14. Haufe, S., Meinecke, F., Gorgen, K., Dahne, S., Haynes, J.D., Blankertz, B., Biessmann, F.: Parameter interpretation, regularization and source localization in multivariate linear models. In: 2014 International Workshop on Pattern Recognition in Neuroimaging, pp. 1–4. IEEE (2014)
15. Henson, R.N., Wakeman, D.G., Litvak, V., Friston, K.J.: A parametric empirical Bayesian framework for the EEG/MEG inverse problem: generative models for multi-subject and multi-modal integration. Front. Hum. Neurosci. **5**, 76 (2011)
16. Kia, S.M.: Mass-univariate hypothesis testing on MEEG data using cross-validation. Master's thesis, University of Trento (2013)
17. Maris, E.: Statistical testing in electrophysiological studies. Psychophysiology **49**(4), 549–565 (2012)

18. Maris, E., Oostenveld, R.: Nonparametric statistical testing of EEG- and MEG-data. J. Neurosci. Methods **164**(1), 177–190 (2007)
19. Olivetti, E., Kia, S.M., Avesani, P.: MEG decoding across subjects. In: 2014 International Workshop on Pattern Recognition in Neuroimaging (2014)
20. Olivetti, E., Kia, S.M., Avesani, P.: Sensor-level maps with the kernel two-sample test. In: 2014 International Workshop on Pattern Recognition in Neuroimaging, pp. 1–4. IEEE (2014)
21. Olivetti, E., Mognon, A., Greiner, S., Avesani, P.: Brain decoding: biases in error estimation. In: 2010 First Workshop on Brain Decoding: Pattern Recognition Challenges in Neuroimaging (WBD), pp. 40–43. IEEE (2010)
22. Rasmussen, P.M., Hansen, L.K., Madsen, K.H., Churchill, N.W., Strother, S.C.: Model sparsity and brain pattern interpretation of classification models in neuroimaging. Pattern Recogn. **45**(6), 2085–2100 (2012)
23. Rish, I., Cecchi, G.A., Lozano, A., Niculescu-Mizil, A.: Practical Applications of Sparse Modeling. MIT Press, Cambridge (2014)
24. Strother, S.C., Rasmussen, P.M., Churchill, N.W., Hansen, K.: Stability and Reproducibility in fMRI Analysis. Springer, New York (2014)
25. Tibshirani, R., Saunders, M., Rosset, S., Zhu, J., Knight, K.: Sparsity and smoothness via the fused lasso. J. Roy. Stat. Soc.: Ser. B (Stat. Methodol.) **67**(1), 91–108 (2005)
26. Valverde-Albacete, F.J., Peláez-Moreno, C.: 100% classification accuracy considered harmful: the normalized information transfer factor explains the accuracy paradox. PLoS ONE **9**(1), e84217 (2014)
27. Varoquaux, G., Gramfort, A., Thirion, B.: Small-sample brain mapping: sparse recovery on spatially correlated designs with randomization and clustering. arXiv preprint arXiv:1206.6447 (2012)
28. Xing, E.P., Kolar, M., Kim, S., Chen, X.: High-dimensional sparse structured input-output models, with applications to GWAS. In: Practical Applications of Sparse Modeling, p. 37 (2014)
29. Yuan, M., Lin, Y.: Model selection and estimation in regression with grouped variables. J. Roy. Stat. Soc.: Ser. B (Stat. Methodol.) **68**(1), 49–67 (2006)
30. Zhang, T., Ghanem, B., Liu, S., Ahuja, N.: Robust visual tracking via structured multi-task sparse learning. Int. J. Comput. Vis. **101**(2), 367–383 (2013)
31. Zhou, J., Chen, J., Ye, J.: MALSAR: Multi-task Learning via Structural Regularization. Arizona State University (2011). http://www.public.asu.edu/jye02/Software/MALSAR
32. Zou, H., Hastie, T.: Regularization and variable selection via the elastic net. J. Roy. Stat. Soc.: Ser. B **67**(2), 301–320 (2005)

The New Graph Kernels on Connectivity Networks for Identification of MCI

Biao Jie[1,2], Xi Jiang[3], Chen Zu[1], and Daoqiang Zhang[1(✉)]

[1] Department of Computer Science and Engineering,
Nanjing University of Aeronautics and Astronautics, Nanjing 210016, China
{jbiao,chenzu,dqzhang}@nuaa.edu.cn
[2] Department of Computer Science and Technology,
Anhui Normal University, Wuhu 241003, China
[3] Department of Computer Science, 415 Boyd Graduate Studies Research Center,
The University of Georgia, Athens, GA 30602-7404, USA
superjx2318@gmail.com

Abstract. Brain connectivity networks have been applied recently to brain disease diagnosis and classification. Especially for both functional and structural connectivity interaction, graph theoretical analysis provided a new measure for human brain organization in vivo, with one fundamental challenge that is how to define the similarity between a pair of graphs. As one kind of similarity measure for graphs, graph kernels have been widely studied and applied in the literature. However, few works exploit to construct graph kernels for brain connectivity networks, where each node corresponds a unique EEG electrode or regions of interest(ROI). Accordingly, in this paper, we construct a new graph kernel for brain connectivity networks, which takes into account the inherent characteristic of nodes and captures the local topological properties of brain connectivity networks. To validate our method, we have performed extensive evaluation on a real mild cognitive impairment (MCI) dataset with the baseline functional magnetic resonance imaging (fMRI) data from Alzheimers disease Neuroimaging Initiative (ADNI) database. Our experimental results demonstrate the efficacy of the proposed method.

1 Introduction

As a neurodegenerative disorder, Alzheimer's disease (AD) is the most common form of dementia in elderly population worldwide. It leads to substantial and progressive neuron damage that is irreversible, which eventually causes death. As a prodromal stage of AD, Mild cognitive impairment (MCI) has gained a great deal of attention recently, because disease-modifying therapies for patients at the early stage of AD development will have a much better effect in slowing down the disease progression and helping preserve some cognitive functions of the brain. Thus, the accurate diagnosis of MCI is very important for possible early treatment and possible delay of the AD progression.

In the context of AD and MCI as well as other brain disorders, numerous studies have suggested that the neurodegenerative diseases such as AD and

© Springer International Publishing AG 2016
I. Rish et al. (Eds.): MLINI 2014, LNAI 9444, pp. 12–20, 2016.
DOI: 10.1007/978-3-319-45174-9_2

MCI are related to a large-scale, highly connected functional network, rather than solely one single isolated region [1–4]. Graph theoretical analysis provides a new way for exploring the association between brain functional deficits and the underlying structural disruption related to brain disorders [5–7]. However, different from traditional data in feature spaces, graph (i.e., network) data is not represented as feature vector, which raise one fundamental challenge for graph data that is how to measure the similarity between a pair of graphs. Motivated by this challenge, computing the similarity of graphs has attracted much attention in the last decade. Among all kinds of methods, kernel methods [8] offer a natural framework to study this question. In the literature, graph kernels, i.e., the kernel constructed on graphs, have been proposed and used in diverse fields. However, existing graph kernels may fail to compare a pair of connectivity networks since they don't consider the inherent characteristics of connectivity networks, such as: (1) the uniqueness of each node of brain connectivity network. That is, each node in connectivity network corresponds a unique EEG electrode or region of interest (ROI). Also, there is one-to-one correspondence between same node across different connectivity networks; (2) local topological properties of connectivity networks, which is very important for measuring the similarity between two connectivity networks. To the best of our knowledge, few work exploit to construct graph kernels on brain connectivity networks. See the related works for detail in the next section.

Accordingly, in this paper, motivated by the recent work in [9], we proposed a new graph kernel on brain connectivity networks, which takes into account the inherent characteristic of nodes and captures the local topological properties of connectivity network. We evaluate our proposed method on 149 subjects with the baseline Resting State fMRI (rs-fMRI) data from ADNI database (www.loni. ucla.edu/ADNI), which includes 99 MCI patients and 50 normal controls. The experiment results demonstrate the efficacy of our proposed method.

1.1   Related Works

Informally, a kernel is a function that measures the similarity between a pair of data points. Mathematically, it corresponds to an inner product in a reproducing kernel Hilbert space [8]. Once a kernel is defined, many learning algorithms such as support vector machines (SVM) can be applied. To compare the similarity between two graphs, graph kernels have been proposed and used in diverse fields including image classification [10], protein function prediction [11]. Existing graph kernels can be roughly divided into two categories: (1) kernels defined on unlabeled graphs where each node has no distinct identification except through their interconnectivity [9,12,13]; (2) kernels defined on labeled graphs where each node is assigned a label [14–17].

In the first category, graph kernels defined on unlabeled graphs dont take into account the labeled information of each node and thus may fail to compute the similarity between a pair of labeled graphs (e.g., brain connectivity networks). In the second category, some of graph kernels are infeasible on connectivity networks because of their computation complexity, such as graph kernels in [14].

Also, some of graph kernels are not fitting for computing on connectivity networks. For example, the graph kernels in [18] mainly compared graphs with edge labels, while that in [17] are to compare graphs with continuous-valued node labels. At the same time, some graph kernels [15,16] are constructed based on Weisfeiler-Lehman test of graph isomorphism. However, note that there doesn't have problem of isomorphism between brain connectivity networks when considering the uniqueness information of each node (i.e., two connectivity networks are the same or difference).

2   Our Proposed Graph Kernel

In this section, we will first briefly introduce the existing graph kernel [9], and then present our proposed graph kernel defined on brain connectivity networks.

Given a graph G (denoted by the matrix $A \in R^{m \times m}$) and a number l, where m is the number of nodes in G. To effectively represent a graph, Shrivastava [9] defined a symmetric positive semi-definite matrix $C^G \in R^{l \times l}$ as:

$$C^G(i,j) = cov(\frac{mA^i e}{\|A^i e\|_1}, \frac{mA^j e}{\|A^j e\|_1}) \tag{1}$$

where cov denotes the covariance between two vectors, e denotes the vector of all 1s, $A^i e$ denotes the i-th power iteration of matrix A on a given starting vector e. (Shrivastava 2013) argued that matrix C^G can capture critical information of the underlying graph and own many good properties, such as graph invariant (i.e., isomorphic graphs have the same representation). Furthermore, based on this new mathematical representation of graphs, (Shrivastava 2013) defined an effective graph kernel on a pair of graph G and H as the follows

$$k(G,H) = \exp(-\frac{1}{2}\log(|\Sigma|/\sqrt{(|C^G||C^H|)})) \tag{2}$$

where $|\cdot|$ denotes the determinant and $\Sigma = (C^G + C^H)/2$.

However, the above-mentioned graph kernels also lack of consideration of two important issues of connectivity networks as we discussed at the previous section. To address that problem, we construct a new graph kernel on connectivity networks.

Denote $G, H \in R^{N \times N}$ as a pair of connectivity networks and given a number h. To reflect the local multi-level topology of connectivity network, we construct N groups of sub-networks. Specifically, for connectivity network G, we construct one group of sub-network on each node i, i.e., $G_i^h = G_i^j = (V_i^j, E_i^j)_{j=1,2,\cdots,h}$, where G_i^j denote a sub-network with a set of nodes V_i^j and a set of edges E_i^j. Here, V_i^j is consist of node i and those nodes that their short-path to the node i is less than or equal to j, and E_i^j includes those edges (i.e., connections) occurred in G. So, we can obtain N groups of sub-networks, i.e., $G = \{G_1^h, G_2^h, G_N^h\}$, where N is the number of nodes. Then, for connectivity network H, we repeat the same process, and also obtain N groups of sub-networks, i.e., $H = \{H_1^h, H_2^h, H_N^h\}$ with

$H_i^h = H_i^j = (V'^j_i, E'^j_i)_{j=1,2,\cdots,h}$. Finally, we can define the kernel on connectivity networks G and H by measuring the similarity between a pair of groups of sub-network from the same node, i.e.,

$$k(G,H) = \frac{1}{Nh} \sum_{i=1}^{N} \sum_{j=1}^{h} \exp\left(-\frac{1}{2}\log\left(|\Sigma_i^j|/\sqrt{|C^{G_i^j}||C^{H_i^j}|}\right)\right) \tag{3}$$

Here, $|\cdot|$ denotes the determinant, $\Sigma_i^j = (C^{G_i^j} + C^{H_i^j})/2$, $C^{G_i^j}$ and $C^{H_i^j}$ are two matrices defined on sub-networks G_i^j and H_i^j by Eq. 1.

Theorem 1. *The kernel as defined in* (3) *is a positive valid kernel.*

It is worth noting that (1) on each node we construct a group of sub-networks which reflects the local multi-level topological properties of connectivity network, and the scale of sub-networks is decided by the value of h. Here, multi-level denotes the sub-network with larger value of j will contain much more nodes and edges, and $G_i^j \subseteq G_i^s$ if $j < s$. In practice, the value of h can be decided via inner cross-validation on training subjects; (2) the kernel defined in Eq. (3) compute the similarity on each pair of groups of sub-networks from the same node across different subjects. Therefore, different with graph kernel in [9], our graph kernel takes into account the uniqueness of nodes and one-to-one correspondence between nodes across different subjects, and captures the local topological properties of connectivity network.

3   Experimental Setup

3.1   Subjects and Data Preprocessing

The dataset used in our study is downloaded from the ADNI database, which includes 99 MCI patients (56 EMCI and 43 LMCI) and 50 normal controls (NC), with each subject of MCI or NC being scanned by fMRI. All rs-fMRI data were acquired on 3.0 Tesla Philips scanners (varied models/systems) at multiple sites. There is a range for imaging resolution in X and Y dimensions, which is from 2.29 mm to 3.31 mm and the slice thickness is 3.31 mm. TE (echo time) for all subjects is 30 ms and TR (repetition time) is from 2.2 s to 3.1 s.

The pre-processing steps of the Resting state fMRI (R-fMRI) data include brain skull removal, slice time correction, motion correction, spatial smoothing, and temporal pre-whitening. The pre-processing steps of the T1-weighted data included brain skull removal and tissue segmentation into gray matter (GM), white matter (WM), and cerebrospinal fluid (CSF). The pre-processed T1 image was then co-registered to the first volume of pre-processed R-fMRI data of the same subject and the BOLD signals in GM were merely extracted and adopted to avoid the relatively high proportion of noise caused by the cardiac and respiratory cycles in WM and ventricle [19]. Finally, The brain space of fMRI images of each subject was then parcellated into 90 ROIs based on the Automated Anatomical Labeling (AAL) template [20]. The mean R-fMRI time series of each individual ROI was calculated by averaging the GM-masked BOLD signals among all voxels within the specific ROI. For each subject, a functional connectivity network was constructed with the vertices of network corresponding to the ROIs and the weight of edges corresponding to the Pearson correlation coefficients. Fisher's r-to-z transformation was applied on the elements of the functional connectivity network to improve the normality of the correlation coefficients.

3.2   Classification

Because the functional connectivity networks are intrinsically weighted graphs as well as fully connected, to reflect the multi-level topological properties of connectivity networks, we first simultaneously threshold the connectivity network with multiple different predefined values (in the experiment, for simplicity, we adopted 5 thresholds, i.e., T= [0.30, 0.35, 0.40, 0.45, 0.50]). Here, we select multiple thresholds instead of single threshold, because the connectivity networks with different thresholds may represent different level of topological properties (i.e., the thresholded connectivity networks with larger threshold often preserve fewer connections and thus are sparser in connection), and these properties may be complementary to each other in improving the classification performance. Then, we compute the graph kernels discussed in previous section on each thresholded connectivity network across different subjects. Finally, we adopt the multi-kernel SVM technique used in [21] for final classification.

4   Experimental Results

4.1   Classification Performance

In our experiments, two binary classifiers, i.e., MCI vs. NC, and EMCI vs. LMCI, are built, respectively. We evaluate the classification performance using the leave-one-out (LOO) cross-validation with a SVM classifier (the parameter parameter). We evaluated the performance of different methods by measuring the classification accuracy, sensitivity, specificity, and the area under receiver operating characteristic (ROC) curve (AUC).

We compare our kernels to state-of-the-art kernels, selected so as to represent three major groups of graph kernels on sub-trees, shortest paths and edges respectively. Those graph kernels belong to Weisfeiler-Lehman graph kernel framework proposed in [15] (denoted as WL-subtree, WL-shortestpath and WL-edge, respectively). Bedsides, we also compared the ego-network-based graph kernels proposed by Shrivastava [9] (denoted as Ego-net) and shortest-path-based kernels proposed in [12] (denoted as Shortest-path). Also, we directly converted the connectivity network (matrix) into a vector, and a feature selection method based on Lasso was performed, and a linear SVM was used to classify the MCI patients from NC (denoted as Vec). Classification results of all methods are summarized in Table 1. For comparison, in Table 2, we also give the classification accuracy of different methods using the single thresholded connectivity

Table 1. Classification performances of different methods

Kernels	MCI vs. NC				EMCI vs. LMCI			
	ACC	SEN	SPE	AUC	ACC	SEN	SPE	AUC
Vec	67.1	91.9	18.0	0.58	46.5	39.5	51.8	0.50
Ego-net	71.8	98.0	20.0	0.60	49.5	87.5	0.0	0.50
Shortest-path	69.8	84.8	40.0	0.60	55.6	67.9	39.5	0.56
WL-edge	73.2	85.9	48.0	0.72	60.6	64.3	55.8	0.61
WL-subtree	76.5	99.0	32.0	0.72	63.6	73.2	51.2	0.62
WL-Shortestpath	73.2	84.8	50.0	0.70	63.6	69.6	55.8	0.59
Proposed	82.6	99.0	50.0	0.80	67.7	83.9	46.5	0.70

Table 2. Classification accuracy of different methods on single thresholded connectivity network

Kernels	MCI vs. NC						EMCI vs. LMCI					
	T1	T2	T3	T4	T5	Combined	T1	T2	T3	T4	T5	Combined
Ego-net	68.5	68.5	69.8	68.5	65.1	71.8	45.5	46.5	44.4	44.4	49.5	49.5
Shortest-path	51.7	59.7	68.5	61.1	64.4	69.8	50.5	49.5	45.5	44.4	46.5	55.6
WL-edge	51.7	59.7	68.5	61.1	64.4	73.2	55.6	52.5	60.6	54.5	52.5	60.6
WL-subtree	67.8	65.8	69.1	70.5	69.1	76.5	55.6	59.6	49.5	52.5	48.5	63.6
WL-Shortestpath	53.7	61.1	67.1	63.1	68.5	73.2	55.6	61.6	60.6	56.6	51.5	63.6
Proposed	73.2	72.5	71.8	71.8	71.1	82.6	59.6	61.6	58.6	58.6	61.6	67.7

networks. As shown in Table 1, the proposed method significantly outperforms the other methods on both classification tasks. Specifically, the proposed method yields a classification accuracy of 82.6 % and 67.7 % for MCI vs. NC and EMCI vs. LMCI classification, respectively, whiloposed graph kee the best classification accuracy of other methods are 76.5 % and 63.6 %, respectively. Also the AUC values of proposed method are 0.80 and 0.70 for both classification tasks, which indicates excellent diagnostic power. Besides, Table 2 show that (1) the combination of multiple thresholded connectivity networks performed significantly better than using any single thresholded connectivity network alone, and (2) the performance of proposed graph kernel on each threholded connectivity network is much batter than that of the state-of-the-art graph kernels, which again shows the efficacy of the prrnel.

4.2   The Discriminative Regions

In this subsection, we further investigate the discriminative power of each ROI using proposed graph kernels. Specifically, for each thresholded connectivity network, we first construct a group of subnetworks on each node, and compute the graph kernels on each group of subnetworks across different subjects according to

Eq. (3). Note that each group of subnetworks reflects the local topological properties of a ROI. Then, we compute the classification accuracy of each ROI with SVM classifier using LOO cross-validation strategy, and rank the ROIs according to their classification accuracy and select the top 10 ROIs with the highest classification accuracy. Figure 1 shows those ROIs that are selected from all thresholded networks. The result shows that most of the selected regions, including hippocampus, cingulate, parahippocampal gyrus, amygdala, heschl gyrus, temporal gyrus and temporal pole, are consistent with the previous studies by using group comparison method [22–24].

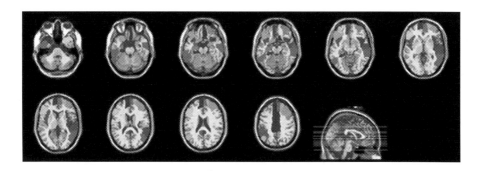

Fig. 1. Top selected ROIs

5   Conclusion

The similarity computation on graph is a fundamental challenge problem in graph-based data analysis. In this paper, we have developed a new graph kernel for measuring the similarity of connectivity networks. Different from the existing graph kernels, our graph kernels take the inherent characteristic of nodes and the local topological properties of connectivity networks into the similarity computation. Series of experiments on real MCI dataset show the efficacy of our proposed method.

Acknowledegment. This work was supported in part by National Natural Science Foundation of China (Nos. 61422204, 61473149), the Jiangsu Natural Science Foundation for Distinguished Young Scholar (No. BK20130034), the NUAA Fundamental Research Funds (No. NE2013105), Natural Science Foundation of Anhui Province (No. 1508085MF125), the Open Projects Program of National Laboratory of Pattern Recognition (No. 201407361).

References

1. Xie, T., He, Y.: Mapping the Alzheimer's brain with connectomics. Front Psychiatry **2**, 77 (2011)
2. Wang, J., Zuo, X., Dai, Z., Xia, M., Zhao, Z., Zhao, X., Jia, J., Han, Y., He, Y.: Disrupted functional brain connectome in individuals at risk for Alzheimer's disease. Biol. Psychiatry **73**, 472–481 (2012)
3. Bai, F., Shu, N., Yuan, Y.G., Shi, Y.M., Yu, H., Wu, D., Wang, J.H., Xia, M.R., He, Y., Zhang, Z.J.: Topologically convergent and divergent structural connectivity patterns between patients with remitted geriatric depression and amnestic mild cognitive impairment. J. Neurosci. **32**, 4307–4318 (2012)
4. Pievani, M., Agosta, F., Galluzzi, S., Filippi, M., Frisoni, G.B.: Functional networks connectivity in patients with Alzheimer's disease and mild cognitive impairment. J. Neurol. **258**, 170–170 (2011)
5. Sporns, O., Tononi, G., Kotter, R.: The human connectome: a structural description of the human brain. PLoS Comput. Biol. **1**, 245–251 (2005)
6. Kaiser, M.: A tutorial in connectome analysis: topological and spatial features of brain networks. Neuroimage **57**, 892–907 (2011)
7. Wee, C.Y., Yap, P.T., Li, W., Denny, K., Browndyke, J.N., Potter, G.G., Welsh-Bohmer, K.A., Wang, L., Shen, D.: Enriched white matter connectivity networks for accurate identification of MCI patients. Neuroimage **54**, 1812–1822 (2011)
8. Scholkopf, B., Smola, A.: Learning with Kernels. The MIT Press, Cambridge (2002)
9. Shrivastava, A., Li, P.: A new mathematical space for social networks. In: Frontiers of Network Analysis: Methods, Models, and Applications, NIPS Workshop, pp. 1–7. MIT Press (2013)
10. Camps-Valls, G., Shervashidze, N., Borgwardt, K.M.: Spatio-spectral remote sensing image classification with graph kernels. IEEE Geosci. Remote Sens. Lett. **7**, 741–745 (2010)
11. Zhang, Y., Lin, H., Yang, Z., Li, Y.: Neighborhood hash graph kernel for protein-protein interaction extraction. J. Biomed. Inform. **44**, 1086–1092 (2011)
12. Borgwardt, K.M., Kriegel, H.P.: Shortest-path kernels on graphs. In: Fifth IEEE International Conference on Data Mining, pp. 74–81 (2005)
13. Johansson, F.D., Jethava, V., Dubhashi, D., Bhattacharyya, C.: Global graph kernels using geometric embedding. In: Proceedings of the 31st International Conference on Machine Learning, vol. 23, pp. 1–9 (2014)
14. Gärtner, T., Flach, P.A., Wrobel, S.: On graph kernels: hardness results and efficient alternatives. In: Schölkopf, B., Warmuth, M.K. (eds.) COLT/Kernel 2003. LNCS (LNAI), vol. 2777, pp. 129–143. Springer, Heidelberg (2003)
15. Shervashidze, N., Schweitzer, P., van Leeuwen, E.J., Mehlhorn, K., Borgwardt, K.M.: Weisfeiler-Lehman graph kernels. J. Mach. Learn. Res. **12**, 2539–2561 (2011)
16. Shervashidze, N., Borgwardt, K.M.: Fast subtree kernels on graphs. In: Advances in Neural Information Processing Systems, vol. 22, pp. 1660–1668 (2009)
17. Feragen, A., Kasenburg, N., Petersen, J., de Bruijne, M., Borgwardt, K.: Scalable kernels for graphs with continuous attributes. In: Advances in Neural Information Processing Systems, pp. 216–224 (2013)
18. Vishwanathan, S.V.N., Schraudolph, N.N., Kondor, R., Borgwardt, K.M.: Graph kernels. J. Mach. Learn. Res. **11**, 1201–1242 (2010)

19. Van Dijk, K.R., Hedden, T., Venkataraman, A., Evans, K.C., Lazar, S.W., Buckner, R.L.: Intrinsic functional connectivity as a tool for human connectomics: theory, properties, and optimization. J. Neurophysiol. **103**, 297–321 (2010)

20. Tzourio-Mazoyer, N., Landeau, B., Papathanassiou, D., Crivello, F., Etard, O., Delcroix, N., Mazoyer, B., Joliot, M.: Automated anatomical labeling of activations in SPM using a macroscopic anatomical parcellation of the MNI MRI single-subject brain. Neuroimage **15**, 273–289 (2002)

21. Zhang, D., Wang, Y., Zhou, L., Yuan, H., Shen, D.: Multimodal classification of Alzheimer's disease and mild cognitive impairment. Neuroimage **55**, 856–867 (2011)

22. Lenzi, D., Serra, L., Perri, R., Pantano, P., Lenzi, G.L., Paulesu, E., Caltagirone, C., Bozzali, M., Macaluso, E.: Single domain amnestic MCI: a multiple cognitive domains fMRI investigation. Neurobiol. Aging **32**, 1542–1557 (2011)

23. Han, Y., Wang, J., Zhao, Z., Min, B., Lu, J., Li, K., He, Y., Jia, J.: Frequency-dependent changes in the amplitude of low-frequency fluctuations in amnestic mild cognitive impairment: a resting-state fMRI study. Neuroimage **55**, 287–295 (2011)

24. Nobili, F., Salmaso, D., Morbelli, S., Girtler, N., Piccardo, A., Brugnolo, A., Dessi, B., Larsson, S.A., Rodriguez, G., Pagani, M.: Principal component analysis of FDG PET in amnestic MCI. Eur. J. Nucl. Med. Mol. **I**(35), 2191–2202 (2008)

Mapping Tractography Across Subjects

Thien Bao Nguyen[1,2(✉)], Emanuele Olivetti[2,3], and Paolo Avesani[2,3]

[1] Faculty of Information Technology, University of Technology and Education,
HoChiMinh City, HoChiMinh, Vietnam
[2] NeuroInformatics Laboratory (NILab), Bruno Kessler Foundation, Trento, Italy
baont@fit.hcmute.edu.vn
http://nilab.fbk.eu
[3] Center for Mind and Brain Sciences (CIMeC), University of Trento, Trento, Italy

Abstract. Diffusion magnetic resonance imaging (dMRI) and tractography provide means to study the anatomical structures within the white matter of the brain. When studying tractography data across subjects, it is usually necessary to align, i.e. to register, tractographies together. This registration step is most often performed by applying the transformation resulting from the registration of other volumetric images (T1, FA). In contrast with registration methods that *transform* tractographies, in this work, we try to find which streamline in one tractography correspond to which streamline in the other tractography, without any transformation. In other words, we try to find a *mapping* between the tractographies. We propose a graph-based solution for the tractography mapping problem and we explain similarities and differences with the related well-known graph matching problem. Specifically, we define a loss function based on the pairwise streamline distance and reformulate the mapping problem as combinatorial optimization of that loss function. We show preliminary promising results where we compare the proposed method, implemented with simulated annealing, against a standard registration techniques in a task of segmentation of the corticospinal tract.

1 Introduction

Diffusion magnetic resonance imaging (dMRI) [1] is a modality that provides non-invasive images of the white matter of the brain. DMRI measures the local the diffusion process of the water molecules in each voxel. That process represents structural information of neuronal axons. From dMRI data, tracking algorithms [9,14] allow to reconstruct the $3D$ pathways of axons within the white matter of the brain as a set of streamlines, called tractography. A *streamline* is a 3D polyline representing thousands of neuronal axons in that region of the brain, and a *tractography* is a large set streamlines, usually $\approx 3 \times 10^5$.

Current neuroscientific analyses of white matter tractography data are limited to qualitative intra-subject comparisons. Thus, it is quite difficult to use the information for direct inter-subject comparisons [2,5]. This leads to the

P. Avesani—The research was funded by the Autonomous Province of Trento, Call "Grandi Progetti 2012", project "Characterizing and improving brain mechanisms of attention - ATTEND".

© Springer International Publishing AG 2016
I. Rish et al. (Eds.): MLINI 2014, LNAI 9444, pp. 21–28, 2016.
DOI: 10.1007/978-3-319-45174-9_3

need of initial alignment, or registration, of tractographies via some methods before doing further study. Registration is most often performed by applying the transformation resulting from the registration of other images, such as $T1$ or fractional anisotropy (FA), to tractography [5,6,12]. Recently, [10] proposed group-wise registration using the trajectory data of the streamlines. The idea to work on deterministic tractography rather than other images is quite innovative. And, it may be advantageous to directly align the streamlines because the result would be closely related to the final goal of registration.

Similar to [10], in this work, we explore the idea of working on deterministic tractography rather than other images. However, in contrast to all current tractography registration methods, which are based on rigid or non-rigid shape transformation of one tractography into another, our approach tries to find which streamline of one tractography corresponds to which streamline in the other tractography, without transformations. This correspondence is a *mapping* from one tractography to the other.

In this work we propose to solve the problem of finding the mapping between two tractographies through a graph-based approach similar to that of the well-known graph matching problem [3,13]. In the graph matching problem the aim is to find which node of one graph corresponds to which node of another graph, under the assumption that graphs have the same number of nodes and that the correspondence is one-to-one.

Given a tractography of N streamlines $T = \{s_1, \ldots, s_N\}$ and a distance function d between streamlines, we can create an undirected weighted graph by considering each streamline as a vertex and the edge connecting vertex s_i and s_j as the distance between the two streamlines, $d(s_i, s_j)$. Then, intuitively, the problem of tractography mapping becomes very similar to that of graph matching, but with some key differences. Firstly, the size of the two tractographies/graphs is in general not the same. Global differences in the anatomy of the brains, e.g. different volume, motivates this difference. Secondly, in general there is not a one-to-one correspondence between the streamlines/nodes but a many-to-one correspondence. This is anatomically likely if we consider that a given anatomical structure (*tract*), e.g. the cortico-spinal tract (CST), whose streamlines should have direct correspondence across subjects, may have different thickness, i.e. different number of streamlines. In this case, for example, multiple streamlines of one CST would correspond to a single streamline in the other CST. Because of these differences, it is generally not possible to directly apply efficient graph matching algorithms to the problem of mapping tractographies.

In the following we formally describe the tractography mapping problem starting from the graph matching problem and define the details of the optimization problem to solve. We provide a preliminary algorithmic solution, based on simulated annealing, to minimize the proposed loss function. Then, we apply our proposed solution to a tractography segmentation task in order to compare a standard registration-based method to our proposed method on a fair ground. We conclude the paper with a brief discussion of the preliminary encouraging results.

2 Methods

An undirected weighted graph $G = (V, E)$ of size N is a finite set of vertices $V = \{1, \ldots, N\}$ and edges $E \subset V \times V$. The graph matching problem can be described as follows. Given two graphs G_A to G_B with the *same* number of vertices N, the problem of matching G_A and G_B is to find the correspondence between vertices of G_A and vertices of G_B, which allows to align, or register, G_A and G_B in some optimal way. The correspondence between vertices of G_A and of G_B is defined as a *permutation* P of the N vertices, i.e. there a one-to-one correspondence between the two set of vertices. P is usually represented as a binary $N \times N$ matrix where P_{ij} is equal to 1, if the ith vertex of G_A is matched to the jth vertex of G_B, otherwise 0. Given A and B, i.e. the $N \times N$ adjacency matrices of the two graphs, the quality of the matching is assessed by the discrepancy, or loss, between the graphs after matching as:

$$L(P) = \|A - PBP^\top\|_2 \tag{1}$$

where $\|A\|_2 = \sqrt{\sum_{ij}^N A_{ij}^2}$ is the Frobenius norm. Therefore, the graph matching problem becomes the problem of finding P^* that minimize L over the set of permutation matrices $\mathcal{P}$:

$$P^* = \underset{P \in \mathcal{P}}{\operatorname{argmin}} \|A - PBP^\top\|_2 \tag{2}$$

which is a combinatorial optimization problem. The exact solution to this problem is NP-complete and only approximate solutions are available in practical cases [3,13].

Let $T_A = \{s_1^A, \ldots, s_N^A\}$ and $T_B = \{s_1^B, \ldots, s_M^B\}$, where $s = \{x_1, \ldots, x_{n_s}\}$ is a streamline and $x \in \mathbb{R}^3$, be the tractographies of two subjects. Let d be a distance function between streamlines. We define two graphs G_A and G_B with adjacency matrix $A \in \mathbb{R}^{N \times N}$ and $B \in \mathbb{R}^{M \times M}$ where $A_{ij} = d(s_i^A, s_j^A)$ and $B_{ij} = d(s_i^B, s_j^B)$. Our current choice of d is discussed in Sect. 3, however any common streamline distance from the literature can be used.

The loss function of a *mapping* Q from T_A to T_B is then:

$$L(Q) = \|A - QBQ^\top\|_2 \tag{3}$$

where the mapping Q is a binary $N \times M$ matrix and Q_{ij} is equal to 1, if s_i^A of T_A is mapped to s_j^B of T_B and 0 otherwise. Note that, in general, Q is not a permutation matrix, because multiple streamlines can be mapped into the same one. In order to find the optimal mapping Q^*, we minimize L so that T_B is most similar to T_A:

$$Q^* = \underset{Q \in \mathcal{Q}}{\operatorname{argmin}} \|A - QBQ^\top\|_2 \tag{4}$$

where $\mathcal{Q}$ is the set of all possible mappings. Because in general $N \neq M$ and because Q is a mapping and not just a permutation, the tractography mapping problem has a larger search space than the graph matching problem, i.e. $|\mathcal{Q}| = M^N \gg N! = |\mathcal{P}|$ when $M \approx N$, is much larger than $\mathcal{P}$. As a consequence, the efficient solutions available in the literature of graph matching, e.g. [13], are not applicable, because they heavily rely on the assumptions that we violate here. In Sect. 3 we implemented a simple preliminary solution to the combinatorial optimization problem by means of the Simulated Annealing metaheuristic [8].

2.1 Comparison

In order to compare the proposed method against a standard registration procedure on a fair ground, we cannot rely on the value of the loss function L, because it is defined only in the case of mapping. For this reason, we compared the two approaches on the practical task of automatic tractography segmentation, i.e. finding a given tract of interest in T_B given its segmentation in T_A. Our hypothesis is that reducing L leads to better overlap between tractographies, which is important for practical applications like segmentation. In Sect. 3 we describe an experiment to test this hypothesis and provide the necessary details. Here we introduce the metric that we use for comparing registration and mapping. As proposed in [5], we compare the set of voxels crossed by the streamlines of each tractography after mapping or after registration. As measure of the overlap between T_A and $Q(T_B)$[1], we adopt the Jaccard index:

$$ J(T_A, T_B | Q) = \frac{|T_A \cap Q(T_B)|}{\min\{|T_A|, |Q(T_B)|\}} \tag{5} $$

Note that in the above equation, $|T|$ is the volume computed as number of voxels that any streamline $s \in T$ goes through, and $|T_A \cap Q(T_B)|$ indicates the number of voxels in common between T_A and $Q(T_B)$.

3 Experiments

We designed an experiment to provide empirical evidence that reducing the loss in Eq. 3 is related to an increase of the Jaccard index, i.e. of the overlap between tractographies.

The dataset used for the experiment is based on dMRI data recorded with a $3T$ scanner at Utah Brain Institute, 65 gradients (64 + $b0$); b-value = 1000; anatomical scan ($2 \times 2 \times 2\,\text{mm}^3$). The tractography was reconstructed with the EuDX algorithm [4] using the dipy[2] toolbox. We considered 4 healthy subjects and focused the analysis on the corticospinal tract (CST). CST is a set of streamlines projecting from the lateral medial cortex associated with the motor homunculus. This tract is of main interest for the characterization of neurodegenerative diseases, like the amyotrophic lateral sclerosis (ALS). The CST tracts were segmented by the expert neuroanatomists using a toolbox [11] that supports an interactive selection of streamlines. The size of the segmented tracts is reported in Table 1 (see column *size*).

---

[1] For sake of brevity we denote as $Q(T_B)$ the result of applying mapping Q to T_B.
[2] http://www.dipy.org.

The reference method, against which we compared mapping, is the affine registration of the tractographies in a common MNI space using the voxel-based FLIRT method [7]. The registration is defined as follows: First, FA images were registered to the MNI-FMRIB-58 FA template, then the affine transformation was applied to the tractographies. The Jaccard index computed between the CST_A and CST_B in common space is reported in Table 1 (see column FLIRT).

We then used mapping to compute the same quantity. The first step was encoding the tractographies as graphs, which required to define a distance between streamlines. We refer to the commonly used Mean Average Minimum distance (MAM) [14], based on the Hausdorff distance:

$$d_{MAM}(s, s') = \frac{1}{2}(D(s, s') + D(s', s)) \tag{6}$$

where $D(s, s') = \frac{1}{n_s} \sum_{i=1}^{n_s} d(x_i, s')$, and $d(x, s') = \min_{j=1,\dots,n_{s'}} ||x - x'_j||_2$.

Mapping a tract such as the CST, which usually comprises 10^2 streamlines, to an entire tractography T_B, which usually consist of 10^7 streamlines, is computationally extremely expensive because the space of all possible mappings $\mathcal{Q}$ has size $|T_B|^{|CST|}$. For this reason, we introduced a heuristic to retain some of the streamlines in T_B. The intuitive idea was to define a superset of streamlines of the CST for subject B, denoted CST_B^+. The heuristic is in two steps: first, we computed the medoid s_m of CST_B, and the radius $r = \max\{d(s_m, s_i), \forall s_i \in CST_B\}$. Second, we filtered the streamlines in T_B such that $CST_B^+ = \{s_j \in T_B | d(s_m, s_j) \leq \alpha \cdot r\}$, where $\alpha = 3$. See Table 1, column CST_B?, for the actual sizes of the supersets.

Computing the optimal mapping Q^* requires to solve, even in an approximate way, the minimization problem of Eq. 4. As a preliminary strategy to approximate the optimal mapping Q^*, we implemented the simulated annealing (SA) [8] meta-heuristic, a reference method for combinatorial optimization. SA requires the definition of a function to move from the current state, i.e. the current mapping Q, to a (potentially better) neighbouring one. As transition function we used a stochastic greedy one where, given the current mapping Q, one streamline of CST_A is selected at random and then it is greedily re-mapped to the streamline in CST_B^+ providing the greatest reduction in the loss of Eq. 3. As starting point of the annealing process, we used the 1-nearest neighbour of CST_A with respect to CST_B^+ after the registration of T_A and T_B. We ran the simulated annealing for 1000 iterations, which required a few minutes on a standard computer[3].

The results reported in Fig. 1 show the behaviour of the loss during the optimization process for the mapping of CST_A (subject ID 205), with respect to the tractography of three other subjects (subject IDs 204, 206 and 212). In all cases, as the number of iterations increases, the value of loss function decreases.

---

[3] We are aware that this method of combinatorial optimization can be significantly improved, but we claim that the it was sufficient to do a preliminary investigation of the relation between the loss L and the overlap between tractographies, by means of the Jaccard index.

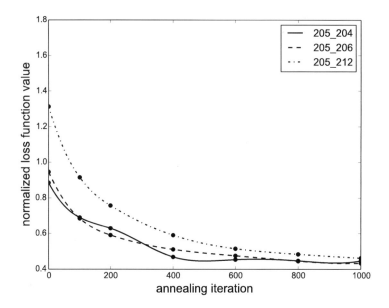

Fig. 1. Plots of the normalized loss ($L_{norm} = \frac{L}{|CST_A|}$) as a function of number of iterations with simulated annealing, when mapping the CST of subject 205 to those of subjects 204, 206 and 212.

In Fig. 2 we show an example of experiment with the outcome of FLIRT registration and mapping which refers to subjects 204 and 206. In subfigure A, the source tract CST_A is shown in blue, in subfigure B the target tract CST_B is show in green and the related superset of streamlines CST_B^+ in red. In subfigure C, the result of FLIRT registration is presented, both with respect to the superset CST_B^+ on the left and with respect to the target tract CST_B on the right. On the right side, it is illustrated the set of streamlines (blue) from the source tract CST_A associated to streamlines of target tract CST_B. The association between streamlines of CST_A and CST_B is computed as nearest neighbour after the FLIRT registration. The ratio between blue and green streamlines represents the portion of target tract correctly detected. On the left side of subfigure C, blue streamlines represents the portion of source tract CST_A not associated to target tract CST_B. In subfigure D, the result of mapping is presented, with the same strategy of presentation of subfigure C. On the right side the visualization shows a greater amount of (blue) streamlines correctly mapped into target tract. Even on the left side the amount of (blue) streamlines erroneously mapped is greater. The sum of blue streamlines on the left and right side represents the portion of streamlines projected from the source to the target. The registration based on FLIRT doesn't preserve after the alignment the same amount of streamlines from the source tract.

In Table 1 are reported the results of the comparison between registration and mapping methods, measured by the Jaccard index. The overlap between CST_A and CST_B provided by FLIRT registration is generally quite poor. This

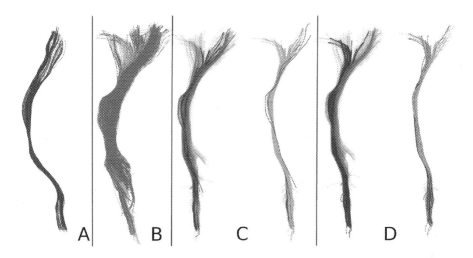

Fig. 2. Example of registration vs. mapping of the corticospinal tract (CST). From the left, the tract to be mapped (subfigure A, CST_A in blue), the second tract with its superset (subfigure B, CST_B in green, CST_B^+ in red), the result of FLIRT affine registration (subfigure C) and of mapping (subfigure D). (Color figure online)

Table 1. Comparison of registration vs. mapping. The subject IDs of CST_A and CST_B are reported in the first two columns. Their sizes, together with that of CST_B^+, are in columns three to five. The last four columns report the overlap between CST_A and CST_B in terms of Jaccard index (higher is better), for FLIRT registration (6th column) and for mapping with simulated annealing at a different number of iterations (SA-0, SA-100, SA-1000 columns).

A	B	size			Jaccard index			
subject ID	subject ID	$\|CST_A\|$	$\|CST_B\|$	$\|CST_B^+\|$	*FLIRT*	*SA-0*	*SA-100*	*SA-1000*
205	204	60	124	682	0.18	0.55	0.52	**0.59**
	206	60	100	550	0.15	0.77	0.81	**0.82**
	212	60	68	374	0.10	0.74	0.77	**0.90**

is partly expected because even after the registration of T_A and T_B, CST_A and CST_B may have a systematic displacement due to the variability of anatomy across subjects. The results of mapping at different iterations of the optimization process shows a remarkable global increase in the Jaccard index and a general trend of improved alignment when more iterations are computed.

4   Discussion and Conclusion

In this work we addressed the challenge of finding an alignment between the tractographies of two subjects. We recast the question as a problem of mapping between two sets of streamlines and we provided the formulation of the corresponding minimization problem. Preliminary results show that this approach is promising despite some limitations. The computational complexity represents a major issue that may prevent to scale up to whole tractography.

References

1. Basser, P.J., Mattiello, J., LeBihan, D.: MR diffusion tensor spectroscopy and imaging. Biophys. J. **66**(1), 259–267 (1994). http://dx.doi.org/10.1016/s0006-3495(94)80775-1
2. Bazin, P.L.L., Ye, C., Bogovic, J.A., Shiee, N., Reich, D.S., Prince, J.L., Pham, D.L.: Direct segmentation of the major white matter tracts in diffusion tensor images. NeuroImage **58**(2), 458–468 (2011). http://dx.doi.org/10.1016/j.neuroimage.2011.06.020
3. Conte, D., Foggia, P., Sansone, C., Vento, M.: Thirty years of graph matching in pattern recognition. Int. J. Patt. Recogn. Artif. Intell. **18**(03), 265–298 (2004). http://dx.doi.org/10.1142/s0218001404003228
4. Garyfallidis, E.: Towards an accurate brain tractography. Ph.D. thesis, University of Cambridge (2012)
5. Golding, D., Tittgemeyer, M., Anwander, A., Douglas, T.: A comparison of methods for the registration of tractographic fibre images. In: Robinson, P., Nel, A. (eds.) Proceedings of the Twenty-Second Annual Symposium of the Pattern Recognition Association of South Africa. pp. 55–59 (2011)
6. Goodlett, C.B., Fletcher, P.T., Gilmore, J.H., Gerig, G.: Group analysis of DTI fiber tract statistics with application to neurodevelopment. NeuroImage **45**(1 Suppl), S133–S142 (2009). http://dx.doi.org/10.1016/j.neuroimage.2008.10.060
7. Jenkinson, M., Smith, S.: A global optimisation method for robust affine registration of brain images. Med. Image Anal. **5**(2), 143–156 (2001). http://view.ncbi.nlm.nih.gov/pubmed/11516708
8. Laarhoven, P.J.M., Aarts, E.H.L. (eds.): Simulated Annealing: Theory and Applications. Kluwer Academic Publishers, Norwell (1987). http://portal.acm.org/citation.cfm?id=59580
9. Mori, S., van Zijl, P.C.M.: Fiber tracking: principles and strategies, a technical review. NMR Biomed. **15**(7–8), 468–480 (2002). http://dx.doi.org/10.1002/nbm.781
10. O'Donnell, L.J., Wells III, W.M., Golby, A.J., Westin, C.-F.: Unbiased groupwise registration of white matter tractography. In: Ayache, N., Delingette, H., Golland, P., Mori, K. (eds.) MICCAI 2012, Part III. LNCS, vol. 7512, pp. 123–130. Springer, Heidelberg (2012)
11. Olivetti, E., Nguyen, T.B., Avesani, P.: Fast Clustering for interactive tractography segmentation. In: The 3rd IEEE Intl Workshop on Pattern Recognition in NeuroImaging (2013). http://dx.doi.org/10.1109/PRNI.2013.20
12. Wang, Y., Gupta, A., Liu, Z., Zhang, H., Escolar, M.L., Gilmore, J.H., Gouttard, S., Fillard, P., Maltbie, E., Gerig, G., Styner, M.: DTI registration in atlas based fiber analysis of infantile Krabbe disease. NeuroImage **55**(4), 1577–1586 (2011)
13. Zaslavskiy, M., Bach, F., Vert, J.P.: A path following algorithm for the graph matching problem. IEEE Trans. Pattern Anal. Mach. Intell. **31**(12), 2227–2242 (2008). http://dx.doi.org/10.1109/tpami.2008.245
14. Zhang, S., Correia, S., Laidlaw, D.H.: Identifying white-matter fiber bundles in DTI data using an automated proximity-based fiber-clustering method. IEEE Trans. Visual. Comput. Graph. **14**(5), 1044–1053 (2008). http://dx.doi.org/10.1109/tvcg.2008.52

Speech

Automated Speech Analysis for Psychosis Evaluation

Facundo Carrillo[1], Natalia Mota[2], Mauro Copelli[3], Sidarta Ribeiro[2],
Mariano Sigman[4], Guillermo Cecchi[5], and Diego Fernandez Slezak[1(✉)]

[1] Laboratorio de Inteligencia Artificial Aplicada,
Facultad de Ciencias Exactas y Naturales, Departamento de Computación,
Universidad de Buenos Aires, Buenos Aires, Argentina
dfslezak@dc.uba.ar
[2] Instituto do Cérebro, Universidade Federal do Rio Grande do Norte, Natal, Brazil
[3] Universidade Federal de Pernambuco, Recife, Brazil
[4] Universidad Torcuato Di Tella, Buenos Aires, Argentina
[5] T.J. Watson Research Center, IBM, Yorktown Heights, NY, USA

Abstract. Psychosis is a mental syndrome associated to loss of contact with reality which may arise in patients with different diseases, such as schizophrenia or bipolar disorder. Symptoms include hallucinations, confused and disturbed thoughts or lack of self-awareness. Recent studies have found that psychotic patients can be objectively screened using graph-theoretical algorithms for speech analysis. This analysis often relies in manually executed tasks such as syntagma generation, text splitting or manual feature selection for classification. To solve this fundamental limitation, we use three fully-automated text analysis tools graph generation methods. In addition, since aspects of psychosis may be manifested in semantic aspects of speech, we also developed a semantic features index based on speech coherence. We show that using this combined approach, classifications obtained from automatic techniques are higher than 85 % in a database of 20 schizophrenic patients, with similar results to previous works. In summary, here we develop and validate a new tool for automated speech processing which includes semantic and structural aspects. The tool performs similar to manual screening procedures providing a new method to complement standard psychometric scales and fostering automated psychiatric diagnosis.

1 Introduction

The way we express ourselves allows us to understand how the brain organizes ideas and concepts, and thus identify and classify the inner organization of thought through the study of speech characteristics. Previous studies have shown that speech is modified under the effects of drug intoxication [9,14]. More recently, speech content analysis provided insight about specific mental-state alterations due to drug ingestion [1].

Mental-state alterations may also reflect psychiatric disorders, such as mania or schizophrenia. The cost to society of depression – one of the most prevalent psiychiatric disorder – is over 40 billion a year in Europe [24]. According to the

© Springer International Publishing AG 2016
I. Rish et al. (Eds.): MLINI 2014, LNAI 9444, pp. 31–39, 2016.
DOI: 10.1007/978-3-319-45174-9_4

National Institutes of Health and the Center for Disease Control, the cost to society in the USA of psychiatric conditions including schizophrenia, depression, alcohol and drug abuse among others surpasses half a trillion dollars per year. These conditions are usually diagnosed, and their treatment monitored, by means of individual oral interviews between psychiatrist and patients.

Computational Psychiatry is a rising new discipline which main objective is the characterization of mental dysfunctions based on the analysis of computational problems resolution [21]. These models seeks to identify traces of cognitive and neural activity, as proposed by Turing and his conception of mental function as information processing modules in a very particular hardware platform: the brain [26]. The massive (and increasingly accelerating) digital availability of thought products in textual format opens a window to study the brain and mind in radically novel ways. We propose that through the use of natural language tools on text and state-of-the-art mathematical approaches we may assess mental states with unprecedented detail and precision.

Mota and colleagues have used graph-based analysis of speech structure to classify manic and schizophrenic patients [23]. Interviews to subjects were manually splitted into syntagmas and manually tagged as dream or waking reports. To avoid manual processing, Mota and colleagues interviewed subjects and requested to *Please report a recent dream* and *Please report your waking activities immediately before that dream*[22]. Graphs were automatically generated from these two different texts. Results showed a high performance of classification between schizophrenic and control subjects ($AUC = 0.94$).

In this work, we hypothesize that speech has the information to sort between classes beyond the asked question. For this, we used both aswsers as one unique report and applied graph analysis without distinction between dream and waking reports. We considered previous structural information – i.e. automated graph analysis – as well as new sintactic information and a complex coherence speech analysis, incorporting semantic features.

Standard graph-analysis captures essentially syntactic and grammatical properties of text. We include the graph generated from POS-tagging of text – which captures more grammatical traces – and coherence analysis [7] which captures the temporal order in discourse. Speech coherence has been vastly studied in different domains, from cognitive science, linguistic to artificial intelligence [12]. This property has different applications: it is useful to quantify the difficulty to comprehend a text [18] helping students to get better texts for their language level, as well a predictor of schizophrenia [7]. In this article, we study the incorporation of coherence analysis parameters to evaluate the increase in classification performance between schizophrenic and control subjects.

2   Methods

2.1   Subjects and Interviews

Fourty subjects participated in the study, 20 of them were diagnosed as schizophrenic and another 20 acted as control. The diagnosis was performed using the standard DSM IV ratings SCID [8]. The schizophrenic subjects were patients of the Hospital Onofre Lopes (UFRN) and Hospital Machado, Natal, Brasil. All subjects were interviewed with the following tasks: *Please report a recent dream* and *Please report your waking activities immediately before that dream.* Their discourse was recorded a blind-conditioned experimenter transcripted the recordings. All subjects signed an informed consent for this study, which was approved by the UFRN Research Ethics Committee (permit #102/06-98244);

For this study, we concatenated the speech for both questions as a unique text. As texts were in Portuguese, text were translated into English using Google Translate.

2.2   Graph Transformations

After transcription of interviews, plain texts were transformed into graphs with three different methods. Naive Graph consisted in splitting the text into words without any transformation. A node was created for every distinct word and two nodes are connected by an edge if they correspond to consecutive words.

Lemma Graph consisted in splitting plain text using the same procedures as in the Naive Graph, but after word lemmatization. With this transformation two different words may have the same lemma (e.g. *love*, *loved* and *loving* have the same lemma: *love*). To lemmatize words, we used the Natural Language Toolkit (NLTK), a Python toolkit for natural language processing [16].

The Part of Speech Graph consisted in changing the words of the plain text with their Part of Speech tags (category of words or lexical items). To perform this, we used the Stanford Tagger [25], building a completely different graph with fewer nodes than other graphs. Figure 1 shows an example of these transformations for a particular sentence.

For each of the three types of graphs, we calculated following graph measures:

- **Nodes:** Number of nodes in the graph
- **Edges:** Number of edges in the graph
- **PE:** The sum of parallel edges in the graph
- **LCC:** Number of nodes in the maximum connected component
- **LSC:** Number of nodes in the maximum Strongly connected component
- **ATD:** Average degree of every node in the graph
- **L1:** Number of self-loops
- **L2:** Number of loops with two nodes
- **L3:** Number of loops with three nodes

2.3   Coherence Algorithm

In this work, we use speech coherence as features to characterize the subject discourse based on the Elvevaag algorithm [7]. The Elvevaag algorithm consists in characterizing the *flow of ideas* based on the psychiatry hypothesis that schizophrenic patients have thought disorders [3].

The coherence analysis method uses Latent Semantic Analysis (LSA) as a vectorial representation of concepts [5,15]. LSA was trained on TASA Corpus, a collection of educational materials compiled by Touchstone Applied Science Associates (TASA). LSA is a high-dimensional associative model that captures similarity between words as a linear representation of the semantic space, by assigning to each word in the lexicon a vector, a distance measure on this space is used to compare words similarity.

The coherence algorithm works as follows: given the speech transcription D, the document is splitted into n phrases S_i, converted into a vectorial representation by replacing each word in the phrase by its corresponding LSA vector, $S_i \rightarrow \{l_{i1}, l_{i2}, ...l_{im}\}$. The phrase vectors are then summarized by taking the mean vector ($L_i = (\sum_{k=1}^{m} l_i)/m$). At this point, the text is a list of vectors with each vector representing a sentence. For each text – i.e. the list of vectors – we define two series: First-Order Coherence (FOC) as the similarity (cosine distance) of consecutive phrase vectors; and Second-Order Cohernece (SOC) as the similarity between phrases separated by another intervening phrase:

$$FOC = [cos(L_1, L_2), cos(L_2, L_3), ..., cos(L_{n-1}, L_n)]$$

$$SOC = [cos(L_1, L_3), cos(L_2, L_4), ..., cos(L_{n-2}, L_n)]$$

Finally, we calculated the mean, median, standard deviation, minimum and max of these series.

2.4   Classifiers

To quantify the effect of the automatic classification of subject groups we used Weka [13], a suite of machine learning methods written in Java, developed at the University of Waikato, New Zealand. We tested the difference between the two approaches: Graphs and Coherence analysis. We ran a selection of classifiers for the three following conditions: (1) using only graph features, (2) using only coherence features and (3) using both, graph and coherence features. For each condition, we took the five classifiers with best performance. We reported the performance of the union of the five classifiers for the three conditions (see Table 1).

The features used in the classification routine were the graph features (9 features per graph) and their normalized version (by the numbers of words in the text), with 18 total features for each type of graph. Thus, The total amount of features for the three types of graphs is 54.

For the coherence analysis we took the two series (FOC and SOC), an calculated many ditribution features: mean, median, standard deviation, minimum and maximum. Again, we generated a normalized version (dividing by the sum of words of the two sentences involved). Moreover, we also analysed the words per sentences series (i.e. word per sentences mean, median, standard deviation, minimum, and max).

3   Results

For each subject, we automatically generated the Naive, Lemma and POS-tag graphs (see Methods section for details) from the original transcript of the interview. Figure 1 shows an example sentence of these transformations. Then, from each graph we measured 18 features. Hence, through this method, we can map each interview to a set 54 features.

Text to Graph transformations example

Original sentence:	*I walked into my house and I found my brother*
Ad-hoc transformation:	*I | walked | into my house | and I | found | my brother*
Naive transformation:	*I | walked | into | my | house | and | I | found | my | brother*
Lemma transformation:	*I | walk | into | my | house | and | I | find| my | brother*
Part of Speech transformation:	*PRP | VBD | IN | PRP$ | NN | CC | PRP | VBD| PRP$ | NN*

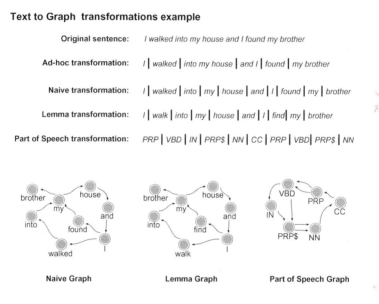

Naive Graph   Lemma Graph   Part of Speech Graph

Fig. 1. Text to graph transformations example.

These 54 features, which are representative of an interview, we trained all Weka classifiers, and computed the performance using a 10-fold cross-validation scheme. First column of Table 1, shows the performance obtained for the selected classifiers, with IB1 (65 %) as the best classifier. This shows the upper-bound of our classification using solely automatically derived syntactic graph metrics.

Next, we calculated the coherence parameters for each interview transcription (see Methods section for details). For each subject, we obtained a set of 30 coherence-related features. Again, we trained all Weka classifiers, and computed the performance using a 10-fold cross-validation scheme (Second column of Table 1). In this case, IB1 – the best performing classifier for graph-only feature – showed slightly decrease of performance, while LWL classifier showed a 82.5 % of performance.

Finally, we combined both sets of features and train all classifiers, and computed the performance using a 10-fold cross-validation scheme. The combination of both groups of features showed the best results, with a 85 % of performance based on LogitBoost classifier [10].

Table 1. Comparison of classifiers using (1) graph features, (2) coherence features and (3) both-types features

Classifiers	Graph Features			Coherence Features			All Features		
	Perf	ROC	F1	Perf	ROC	F1	Perf	ROC	F1
LogitBoost	0.55	0.549	0.549	0.775	0.82	0.774	0.85	0.905	0.85
AdaBoostM1	0.575	0.591	0.573	0.8	0.838	0.799	0.825	0.835	0.825
BayesNet	0.5	0.5	0.33	0.775	0.819	0.774	0.775	0.819	0.774
LWL	0.45	0.513	0.437	0.825	0.733	0.819	0.775	0.723	0.763
DecisionStump	0.45	0.42	0.449	0.775	0.718	0.763	0.775	0.718	0.763
RandomCommittee	0.55	0.626	0.549	0.8	0.828	0.799	0.725	0.76	0.721
OneR	0.65	0.65	0.649	0.625	0.625	0.619	0.625	0.625	0.619
IB1	0.65	0.65	0.65	0.6	0.6	0.599	0.425	0.425	0.416

To evaluate the robustness of these results, we tested the best classifier (LogitBoost) versus a bootstrapping, i.e. a random assignation of labels to subject. In this random assignation, classification methods should be around 50 %, i.e. the chance of assigning the correct label between two possibilities: schizophrenic or control. Figure 2 shows the classification performance of LogitBoost with each of the three groups of features, compared to the random sampling described. Graph features show very similar results to bootstrapping, indicating very low performance to distriminate between categories. Coherence features exhibits better results, with more than 70 % of performance. But combination of both set of features obtain the best results, with more than 82 % of performance, separating from the bootstrapping data.

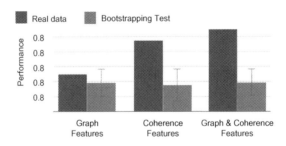

Fig. 2. Performance comparison using graph-only, coherence-only and graph+coherence classification with the LogitBoost classifier (blue bars) and bootstrapping classification (red bars). (Color figure online)

4 Discussion

The field of cognitive sciences has recently developed an approach to amass enormous amounts of data through game-like web applications [11,17,27]. Together with the access to large web repositories of text, and to virtually unlimited computational power, these developments are changing the way we can characterize cognition and human behavior. The massive availability of thought products in digital textual format opens up a new era with a great amount of new possibilities, pushing a profound reformulation of how to study and assess mental states.

Several approaches have been developed in the natural language processing discipline to automatically extract word representation from texts to analyze semantic content [2,4,5,20]. Recent works have shown that these methods may extract regularities in texts reflecting mental/societal activity [6,19].

In this article, we combine discrete mathematics algorithms for graph characterization, with natural language processing techniques to train classifiers that can distinguish interviews from schizophrenic and control subjects. Graph measures concentrate on how text syntax and grammar is used. We incorporate semantic features through the coherence analysis into the classification process. NLP methods – LSA, specifically – relies on corpus training to create a vector space where words (or concepts) may be compared or measured providing a computational measure of semantic content. Here we combine structural and semantic features and evaluated the classification performance to discriminate between these two categories: schizophrenic and control subjects.

We report the results of the classification in these three groups of features (graph features; coherence features and mixed features). Most classifiers show higher performance using coherence features than using graph features. However the combination of both types of features show the best results, showing that there is synergic information between these two aspects of speech.

This new tool presents an automated text processing method which includes coherence analysis. Results complement standard psychometric scales fostering automated psychiatric diagnosis.

Acknowledgments. This research was supported by University of Buenos Aires, CONICET (Argentina) and ANPCyT (Argentina). MS is sponsored by James McDonnell Foundation 21st Century Science Initiative in Understanding Human Cognition. DFS is sponsored by Microsoft Faculty Fellowship.

References

1. Bedi, G., Cecchi, G.A., Slezak, D.F., Carrillo, F., Sigman, M., de Wit, H.: A window into the intoxicated mind? speech as an index of psychoactive drug effects. Neuropsychopharmacology **39**, 2340–2348 (2014)
2. Blei, D., Ng, A., Jordan, M.: Latent dirichlet allocation. J. Mach. Learn. Res. **3**, 993–1022 (2003)
3. Chapman, L.J., Chapman, J.P.: Disordered thought in schizophrenia (1973)
4. Church, K., Hanks, P.: Word association norms, mutual information, and lexicography. Comput. Linguist. **16**(1), 22–29 (1990)
5. Deerwester, S., Dumais, S., Furnas, G., Landauer, T., Harshman, R.: Indexing by latent semantic analysis. J. Am. Soc. Inf. Sci. **41**(6), 391–407 (1990)
6. Diuk, C., Slezak, D., Raskovsky, I., Sigman, M., Cecchi, G.: A quantitative philology of introspection. Front. Integr. Neurosci. **6**, 80 (2012)

7. Elvevåg, B., Foltz, P.W., Weinberger, D.R., Goldberg, T.E.: Quantifying incoherence in speech: an automated methodology and novel application to schizophrenia. Schizophr. Res. **93**(1), 304–316 (2007)
8. First, M.B., Spitzer, R.L., Gibbon, M., Williams, J.B.: Structured Clinical Interview for DSM-IV ® Axis I Disorders (SCID-I), Clinician Version, Administration Booklet. American Psychiatric Pub (2012)
9. Foltin, R.W., Fischman, M.W.: Effects of smoked marijuana on human social behavior in small groups. Pharmacol. Biochem. Behav. **30**(2), 539–541 (1988)
10. Friedman, J., Hastie, T., Tibshirani, R.: Additive logistic regression: a statistical view of boosting (with discussion and a rejoinder by the authors). Ann. Stat. **28**(2), 337–407 (2000)
11. Goldin, A.P., Hermida, M.J., Shalom, D.E., Costa, M.E., Lopez-Rosenfeld, M., Segretin, M.S., Fernández-Slezak, D., Lipina, S.J., Sigman, M.: Far transfer to language and math of a short software-based gaming intervention. Proc. Nat. Acad. Sci. **111**(17), 6443–6448 (2014)
12. Graesser, A.C., McNamara, D.S., Louwerse, M.M., Cai, Z.: Coh-metrix: analysis of text on cohesion and language. Behav. Res. Methods Instrum. Comput. **36**(2), 193–202 (2004)
13. Hall, M., Frank, E., Holmes, G., Pfahringer, B., Reutemann, P., Witten, I.: The weka data mining software: an update. ACM SIGKDD Explor. Newslett. **11**(1), 10–18 (2009)
14. Higgins, S.T., Stitzer, M.L.: Effects of alcohol on speaking in isolated humans. Psychopharmacology **95**(2), 189–194 (1988)
15. Landauer, T., Dumais, S.: A solution to plato's problem: the latent semantic analysis theory of acquisition, induction, and representation of knowledge. Psychol. Rev. **104**(2), 211 (1997)
16. Loper, E., Bird, S.: NLTK: the natural language toolkit. In: Proceedings of the ACL 2002 Workshop on Effective Tools and methodologies for Teaching Natural Language Processing and Computational Linguistics-Volume 1, pp. 63–70. Association for Computational Linguistics (2002)
17. Lopez-Rosenfeld, M., Goldin, A.P., Lipina, S., Sigman, M., Fernandez Slezak, D.: Mate marote: a flexible automated framework for large-scale educational interventions. Comput. Educ. **68**, 307–313 (2013)
18. McNamara, D.S., Louwerse, M.M., Graesser, A.C.: Coh-metrix: automated cohesion and coherence scores to predict text readability and facilitate comprehension. Unpublished Grant Proposal, University of Memphis, Memphis, Tennessee (2002)
19. Michel, J., Shen, Y., Aiden, A., Veres, A., Gray, M., Pickett, J., Hoiberg, D., Clancy, D., Norvig, P., Orwant, J., et al.: Quantitative analysis of culture using millions of digitized books. Science **331**(6014), 176 (2011)
20. Mikolov, T., Sutskever, I., Chen, K., Corrado, G.S., Dean, J.: Distributed representations of words and phrases and their compositionality. In: Burges, C., Bottou, L., Welling, M., Ghahramani, Z., Weinberger, K. (eds.) Advances in Neural Information Processing Systems, vol. 26, pp. 3111–3119. Curran Associates, Inc., New York (2013). http://papers.nips.cc/paper/5021-distributed-representations-of-words-and-phrases-and-their-compositionality.pdf
21. Montague, P., Dolan, R., Friston, K., Dayan, P.: Computational psychiatry. Trends Cogn. Sci. **16**(1), 72–80 (2012)
22. Mota, N.B., Furtado, R., Maia, P.P., Copelli, M., Ribeiro, S.: Graph analysis of dream reports is especially informative about psychosis. Sci. Rep. **4** (2014)

23. Mota, N., Vasconcelos, N., Lemos, N., Pieretti, A., Kinouchi, O., Cecchi, G., Copelli, M., Ribeiro, S.: Speech graphs provide a quantitative measure of thought disorder in psychosis. PLoS ONE **7**(4), e34928 (2012)

24. Sobocki, P., Jönsson, B., Angst, J., Rehnberg, C.: Cost of depression in Europe. J. Ment. Health Policy Econ. **9**(2), 87–98 (2006)

25. Toutanova, K., Klein, D., Manning, C.D., Singer, Y.: Feature-rich part-of-speech tagging with a cyclic dependency network. In: Proceedings of the 2003 Conference of the North American Chapter of the Association for Computational Linguistics on Human Language Technology-Volume 1, pp. 173–180. Association for Computational Linguistics (2003)

26. Turing, A.: Computing machinery and intelligence. Mind **59**(236), 433–460 (1950)

27. Von Ahn, L.: Games with a purpose. Computer **39**(6), 92–94 (2006)

Combining Different Modalities in Classifying Phonological Categories

Shunan Zhao[1(✉)] and Frank Rudzicz[1,2]

[1] University of Toronto, Toronto, Canada
frank@cs.toronto.edu
[2] Toronto Rehabilitation Institute, University of Toronto, Toronto, Canada

Abstract. This paper concerns a new dataset we are collecting combining 3 modalities (EEG, video of the face, and audio) during imagined and vocalized phonemic and single-word prompts. We pre-process the EEG data, compute features for all 3 modalities, and perform binary classification of phonological categories using a combination of these modalities. For example, a deep-belief network obtains accuracies over 90 % on identifying consonants, which is significantly more accurate than two baseline support vector machines. These data may be used generally by the research community to learn multimodal relationships, and to develop silent-speech and brain-computer interfaces.

Keywords: Phonological categories · Electroencephalography · Speech articulation · Deep-belief networks

1   Introduction

Brain-computer interfaces (BCIs) often involve imagining gross motor movements (e.g., of the hands or feet) to move a pointer on-screen or select from banks of characters. However, some recent research has attempted to access language centres in the brain directly. This has included electrocorticography (ECoG) [2,10] and intracranial neurotrophic electrodes [1] to estimate perceived words or complete auditory spectra [12] directly. While invasive methods have high signal-to-noise, they can often only be used in severe cases, due to their complex nature. We are therefore interested in discovering solutions that can be applied more generally.

Suppes *et al.* [16] performed whole-word recognition using using EEG and magnetoencephalography (MEG) data, where participants either silently pronounced words or thought abstractly about their meaning. Porbadnigk *et al.* [13] used an HMM to classify between EEG signals associated with the imagined speech of five words with limited accuracy. The order in which the words were presented significantly affected the results, which were above chance for only one of four modes. Previous attempts to classify EEG signals associated with the imagined pronunciation of phonemes often focussed on vowels [3–5,7], building on previous work by Fujimaki *et al.* [8], who identified event-related

© Springer International Publishing AG 2016
I. Rish et al. (Eds.): MLINI 2014, LNAI 9444, pp. 40–48, 2016.
DOI: 10.1007/978-3-319-45174-9_5

potentials during the imagined pronunciation of /a/. While relevant, these studies did not relate EEG signals to either articulation or acoustics during actual speech production, which is central to our work.

2 Data

2.1 Data Collection

Eight female and eleven male participants (mean age = 27.4, $\sigma = 6.9$, range = 26) were recruited from the University of Toronto. All participants were right-handed (to control for hemispheric differences), had at least some post-secondary education, had no visual, hearing, or motor impairments, and had no history of neurological conditions or drug abuse. Furthermore, 17 of the 19 participants identified North American English as their first language and the remaining two spoke North American English at a fluent level, having learned the language at a mean age of six.

Each study was conducted in an office environment at the Toronto Rehabilitation Institute (part of the University Health Network). Each participant was seated in a chair before a computer monitor. A Microsoft Kinect (v.1.8) sensor was placed above the screen to record facial information and the audio of the participant's speech. For each frame of video, the Kinect's bundled software extracted six 'animation units' (AUs), all on $\mathbb{R}[-1..1]$: *upper lip raiser, jaw lowerer, (lateral) lip stretcher, brow lowerer, lip corner depressor, outer brow raiser*. A research assistant placed an appropriately-sized EEG cap on the participant's head and injected a small amount of gel to improve electrical conductance. We used a 64-channel Neuroscan Quick-cap, where the electrode placement follows the 10–20 system [14]. To control for artifacts arising from eye-movement, we used 4 electrodes placed above and below the left eye and to the lateral side of each eye. All EEG data were recorded using the SynAmps RT amplifier and sampled at 1 kHz. Impedance levels were usually maintained below 10 kΩ.

After EEG setup, the participant was instructed to look at the computer monitor and to move as little as possible. Over the course of 30 to 40 min, individual prompts appeared on the screen one-at-a-time. We used 7 phonemic/syllabic prompts (/iy/, /uw/, /piy/, /tiy/, /diy/, /m/, /n/) and 4 words derived from Kent's list of phonetically-similar pairs (i.e., *pat, pot, knew*, and *gnaw*) [11]. These prompts were chosen to maintain a relatively even number of nasals, plosives, and vowels, as well as voiced and unvoiced phonemes. Each trial consisted of 4 successive states:

1. A 5-second **rest** state, where the participant was instructed to relax and clear their mind of any thoughts.
2. A **stimulus** state, where the prompt text would appear on the screen and its associated auditory utterance was played over the computer speakers. This was followed by a 2-second period in which the participant moved their articulators into position to begin pronouncing the prompt.

3. A 5-second **imagined** speech state, in which the participant imagined speaking the prompt without moving.
4. A **speaking** state, in which the participant spoke the prompt aloud. The Kinect sensor recorded both the audio and facial features during this stage.

Naturally, given the impact of movement on EEG, we expect excessive noise in the **speaking** state EEG. Once the participant has finished speaking, one of the investigators would proceed to the next trial. Each prompt was presented 12 times for a total of 132 trials each. The phonemic/syllabic prompts were first presented followed by the 4 'Kent' words, and the trials were randomly permuted within each of those two sections. After every 40 trials, the participant was given the opportunity to rest. Data from 5 of the 19 participants were discarded due to unattached ground wires and two participants falling asleep during recording. Ethical approval was obtained from both the University of Toronto and the University Health Network, of which Toronto Rehab is a member.

2.2   Pre-processing

EEG data were pre-processed with EEGLAB [6], including removal of ocular artifacts using blind source separation [9]. The data were band-pass filtered between 1 Hz and 50 Hz [5], and the mean values were subtracted from each channel. The EEG data were segmented into different trials, and each trial was further segmented into the 4 states described above. We discarded 16 trials that did not contain facial features from the Kinect.

2.3   Feature Extraction and Selection

For each EEG segment and each non-ocular channel, we window the data to approximately 10 % of the segment, with a 50 % overlap between consecutive windows. We then compute several features over each window, including the mean, median, standard deviation, variance, maximum, minimum, maximum ± minimum, sum, spectral entropy, energy, kurtosis, and skewness. We also compute the mean, maximum, minimum, and the sum and difference of the maximum and minimum for the absolute value of the windowed signal. Furthermore, we compute the first and second derivates of the above features. This results in 1197 features for each channel of the segment, for a total of 65,835 features across the 62 channels. For each audio recording, we measure the same set of features. For the facial data, we measure a subset of the above features for each AU, including the mean, maximum, minimum, median, skewness, and kurtosis. We further compute the first and second derivatives for each AU and measure the same set of features.

Due to the high dimensionality of the feature space, particularly for the EEG features, we rank features by their Pearson correlations with the given classes for each task independently and we select the N features with the highest correlation coefficients, where $N \in [5...100]$. Given the multiple tasks and our cross-validation scheme (see Sect. 3), we perform feature selection on every training set independently.

3   Experiments

We use a subject-independent approach with leave-one-out cross-validation in which each subject's data are tested in turn using models trained with all other data combined. The results therefore may provide more generalizable conclusions than subject-specific models, typical in the literature, which depend on individual, non-transferable models. Our experiments use two types of classifier: a deep-belief network (**DBN**) and support vector machine (SVM) baselines. Two variants of the latter are tested, with different kernels; **SVM-quad** uses a quadratic kernel and **SVM-rbf** uses the radial basis function.

In the DBN, weights w_{ij} between nodes i and j, in different layers, are adjusted at iteration $t+1$ with gradient descent given weights at time t according to

$$\Delta w_{ij}(t+1) = w_{ij}(t) + \eta \frac{\delta \log(P(\mathbf{x}))}{\delta w_{ij}}, \tag{1}$$

After unsupervised training, we set a linear mapping of the output and 'fine tune' the network in a supervised fashion using class predictions. In all cases, we use one hidden layer whose (bottleneck) size is empirically 25 % of the size of the input. We use up to 10 iterations (to avoid overfitting) in the pretraining cycle with a batchsize of $N/4$ (given N observation vectors), a learning rate $\eta = 0.1$, a drop-out rate [15] of 0.5 and the 'cross entropy' objective function $C = -\sum_j d_j \log(p_j)$, empirically chosen, where d_j is the target probability for output j and p_j is the actual probability output of j.

3.1   Classification of Phonological Categories

Our primary goal is to classify between important phonemic and phonological classes given different modalities of data and speech *planning* and *production*. Specifically, we consider five binary classification tasks: vowel-only vs. consonant (**C/V**), presence of nasal (± **Nasal**), presence of bilabial (± **Bilab.**), presence of high-front vowel (±/**iy**/), and presence of high-back vowel (±/**uw**/) using six modalities: **EEG**-only, facial features (**FAC**)-only, audio (**AUD**)-only, EEG and facial features (**EEG+FAC**), EEG and audio features (**EEG+AUD**), and all modalities.

Figures 1 and 2 show the average accuracy (with standard error $\sigma/\sqrt{n}$) of classifying ±/*iy*/ and C/V, respectively, across the three classifiers and for each test subject (given subject-independent models trained on all other data) given empirical $N = 5$ input features. For both tasks, the DBN classifiers obtain between 80 % and 91 % accuracy. Although the SVM-quad classifier obtains significantly better-than-chance accuracy on ±/*uw*/, the SVM classifiers, in general, obtain significantly lower accuracy than the DBNs. As suggested by the high $\sigma/\sqrt{n}$ for the SVM classifiers, this may be largely due to an interaction between the classification tasks and the modalities of the data used. Indeed, Table 1 shows that the average accuracies of the SVM-quad classifier varies greatly across these two dimensions. This is further confirmed by an analysis of variance, which not

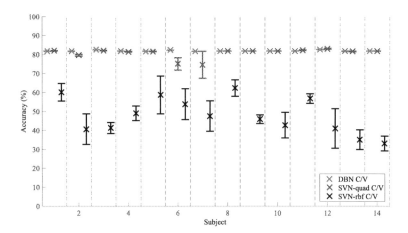

Fig. 1. Average accuracies across models for DBN, SVN-quad, and SVN-rbf classifiers for the C/V task, across subjects. Error bars are $\sigma/\sqrt{n}$.

only shows significant linear effects of each of the classifier, test subject, task, and modality on the accuracy of phonological category classification, but also significant interactions between the task and both of the classifier used and the modality of the data ($p < 0.01$).

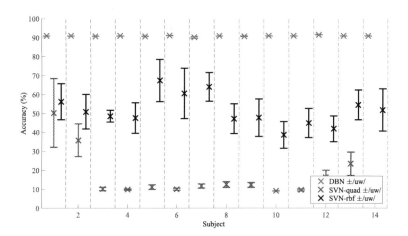

Fig. 2. Average accuracies across models for DBN, SVN-quad, and SVN-rbf classifiers for the $\pm/uw/$ task, across subjects. Error bars are $\sigma/\sqrt{n}$.

3.2   Correlational Analysis

To further investigate which features are most useful across the three different modalities, we compute Pearson's correlation between all pairs of features *across* modality pairs (i.e., EEG and WAV, EEG and FAC, WAV and FAC). These correlation coefficients are computed over all users and all trials. To make these features more interpretable, we restrict ourselves to a smaller feature set consisting mostly of the same features that we computed on the facial data. The results are summarized in Tables 2, 3, and 4. While the correlation coefficients are not particularly high, it is interesting to note that certain features appear significantly more often than others, such as the skewness, variance, and the sum of the first derivative of the signal. We are conducting a more thorough analysis of how different features from the various modalities relate to each other.

Table 1. Average accuracies (%) across modalities and classes given the SVM-quad classifier.

	\multicolumn{5}{c}{Task}				
	C/V	± Nasal	± Bilab	±/iy/	±/uw/
EEG	78.93	37.41	37.47	37.1	9.53
FAC	81.7	36.6	36.44	37.16	12.18
AUD	80.65	42.98	59.81	37.56	22.41
EEG+FAC	81.7	36.48	36.44	37.64	10.26
EEG+AUD	80.65	42.98	59.81	37.87	22.41
ALL	80.7	37.17	56.06	38.98	22.41

Table 2. The 10 highest absolute correlated features for the EEG and WAV modality pair. The parentheses indicate the channel label.

EEG feature	WAV feature	r
(CZ) Min of 2nd derivative	Mean of 6th power	0.33
(P6) Min of 1st derivative	Mean of 6th power	0.33
(AF3) Sum of 1st derivative	Variance	0.31
(AF3) Sum of 1st derivative	Absolute mean	0.30
(FT8) Kurtosis of 2nd derivative	Variance	0.30
(P3) Kurtosis of 1st derivative	Variance	0.30
(AF3) Sum of 1st derivative	Absolute mean of 1st derivative	0.30
(AF3) Sum of 1st derivative	Absolute mean of 2nd derivative	0.30
(AF3) Sum of 1st derivative	std. dev	0.30
(CP1) Median	Max	0.30

Table 3. The 10 highest absolute correlated features for the EEG and FAC modality pair. The parentheses indicate the channel label and AU (AU0: upper lip raiser).

EEG feature	FAC feature	r
(C4) Mean of 6th power (2nd derivative)	(AU0) Mean of 1st derivative	-0.41
(PO7) Mean of 6th power (1st derivative)	(AU0) Mean of 1st derivative	-0.41
(CP4) Variance of 1st derivative	(AU0) Mean of 1st derivative	0.40
(O2) Variance	(AU0) Mean of 1st derivative	0.40
(CP4) Min of 1st derivative	(AU0) Mean of 1st derivative	0.39
(O2) Min	(AU0) Mean of 1st derivative	0.39
(C4) Mean of 6th power (2nd derivative)	(AU0) std. dev. of 2nd derivative	-0.38
(PO7) Mean of 6th power (1st derivative)	(AU0) std. dev. of 2nd derivative	-0.38
(TP8) Kurtosis of 1st derivative	(AU0) Mean of 1st derivative	-0.38
(O1) Kurtosis	(AU0) Mean of 1st derivative	-0.38

Table 4. The 10 highest absolute correlated features for the WAV and FAC modality pair. Here, AU0: upper lip raiser; AU2: lip stretcher; AU3: brow lowerer; AU4: lip corner depressor; AU5: outer brow raiser.

WAV feature	FAC feature	r
Skewness	(AU3) Variance	0.31
Skewness	(AU0) Min of first derivative	0.30
Skewness	(AU0) Sum	0.30
Skewness	(AU0) Sum of 2nd derivative	0.30
Skewness	(AU5) Variance	0.30
Skewness	(AU2) Sum of 2nd derivative	0.30
Skewness	(AU2) Sum	0.29
Skewness	(AU2) Max minus min	0.29
Skewness	(AU2) Min of 1st derivative	0.29
Skewness	(AU4) Sum of 2nd derivative	0.29

4   Discussion

In this paper, we classify between phonological categories in planned and executed speech combining acoustic, facial, and EEG data. Usually such multimodality is only possible with expensive MEG equipment. Instead, we use an affordable (and portable) Kinect sensor and 64-channel EEG cap, which is a much more viable setup for brain-computer interfaces. This data set is also notable in that it combines EEG and physical speech production, which is normally limited due to inherent measurement artifacts. Furthermore, all our reported experiments use leave-one-out cross-validation, so our models are subject-independent and generalizable.

Other preliminary experiments are presented in the companion paper to this work [17], which includes classification of the mental state from EEG, which can achieve up to 95 % with the DBN described here. That task may be involved in triggering silent text entry, for example, especially when comparing resting states from active phonological planning. Future work includes methods to reconstruct acoustic features from EEG, after Pasley *et al.*'s work with invasive methods [12], potentially towards mapping imagined speech to synthetic speech.

Acknowledgements. This research is funded by the Toronto Rehabilitation Institute, the Natural Sciences and Engineering Research Council of Canada (RGPIN 435874), and a grant from the Nuance Foundation. Data collection was assisted by Selvana Morcos, Aaron Marquis, Chaim Katz, and César Márquez-Chin.

References

1. Bartels, J., Andreasen, D., Ehirim, P., Mao, H., Seibert, S., Wright, E.J., Kennedy, P.: Neurotrophic electrode: method of assembly and implantation into human motor speech cortex. J. Neurosci. Methods **174**(2), 168–176 (2008). http://www.sciencedirect.com/science/article/pii/S0165027008003865

2. Blakely, T., Miller, K., Rao, R.P.N., Holmes, M.D., Ojemann, J.: Localization and classification of phonemes using high spatial resolution electrocorticography (ECoG) grids. In: 30th Annual International Conference of the IEEE Engineering in Medicine and Biology Society, EMBS 2008, pp. 4964–4967, August 2008

3. Brigham, K., Kumar, B.: Imagined speech classification with EEG signals for silent communication: a preliminary investigation into synthetic telepathy. In: 2010 4th International Conference on Bioinformatics and Biomedical Engineering (iCBBE), pp. 1–4, June 2010

4. Callan, D.E., Callan, A.M., Honda, K., Masaki, S.: Single-sweep EEG analysis of neural processes underlying perception and production of vowels. Cognit. Brain Res. **10**(1–2), 173–176 (2000). http://www.sciencedirect.com/science/article/pii/S0926641000000252

5. DaSalla, C.S., Kambara, H., Sato, M., Koike, Y.: Single-trial classification of vowel speech imagery using common spatial patterns. Neural Netw. **22**(9), 1334–1339 (2009). http://www.sciencedirect.com/science/article/pii/S0893608009000999, brain-Machine Interface

6. Delorme, A., Makeig, S.: EEGLAB: an open source toolbox for analysis of single-trial EEG dynamics including independent component analysis. J. Neurosci. Methods **134**(1), 9–21 (2004). http://www.sciencedirect.com/science/article/pii/S0165027003003479

7. D'Zmura, M., Deng, S., Lappas, T., Thorpe, S., Srinivasan, R.: Toward EEG sensing of imagined speech. In: Jacko, J.A. (ed.) HCI International 2009, Part I. LNCS, vol. 5610, pp. 40–48. Springer, Heidelberg (2009)

8. Fujimaki, N., Takeuchi, F., Kobayashi, T., Kuriki, S., Hasuo, S.: Event-related potentials in silent speech. Brain Topogr. **6**(4), 259–267 (1994)

9. Gomez-Herrero, G., De Clercq, W., Anwar, H., Kara, O., Egiazarian, K., Van Huffel, S., Van Paesschen, W.: Automatic removal of ocular artifacts in the EEG without an EOG reference channel. In: Proceedings of the 7th Nordic Signal Processing Symposium, NORSIG 2006, pp. 130–133, June 2006

10. Kellis, S., Miller, K., Thomson, K., Brown, R., House, P., Greger, B.: Decoding spoken words using local field potentials recorded from the cortical surface. J. Neural Eng. **7**(5), 1–10 (2010). http://stacks.iop.org/1741-2552/7/i=5/a=056007
11. Kent, R.D., Weismer, G., Kent, J.F., Rosenbek, J.C.: Toward phonetic intelligibility testing in dysarthria. J. Speech Hear. Disord, **54**(4), 482–499 (1989). http://dx.doi.org/10.1044/jshd.5404.482
12. Pasley, B.N., David, S.V., Mesgarani, N., Flinker, A., Shamma, S.A., Crone, N.E., Knight, R.T., Chang, E.F.: Reconstructing speech from human auditory cortex. PLoS ONE **10**(1), 1–13 (2012)
13. Porbadnigk, A., Wester, M., Calliess, J., Schultz, T.: EEG-based speech recognition- impact of temporal effects. In: Encarnao, P., Veloso, A. (eds.) BIOSIGNALS, pp. 376–381. INSTICC Press (2009)
14. Sharbrough, F., Chatrian, G., Lesser, R., Lüders, H., Nuwer, M., Picton, T.: American electroencephalographic society guidelines for standard electrode position nomenclature. J. Clin. Neurophysiol. **8**(2), 200–202 (1991)
15. Srivastava, N., Hinton, G., Krizhevsky, A., Sutskever, I., Salakhutdinov, R.: Dropout: a simple way to prevent neural networks from overfitting. J. Mach. Learn. Res. **15**, 1929–1958 (2014). http://jmlr.org/papers/v15/srivastava14a.html
16. Suppes, P., Lu, Z.L., Han, B.: Brain wave recognition of words. Proc. Nat. Acad. Sci. **94**(26), 14965–14969 (1997). http://www.pnas.org/content/94/26/14965.abstract
17. Zhao, S., Rudzicz, F.: Classifying phonological categories in imagined and articulated speech. In: Proceedings of ICASSP 2015 (2015)

Clinics and Cognition

Label-Alignment-Based Multi-Task Feature Selection for Multimodal Classification of Brain Disease

Chen Zu, Biao Jie, Songcan Chen, and Daoqiang Zhang[✉]

Department of Computer Science and Engineering,
Nanjing University of Aeronautics and Astronautics, Nanjing 210016, China
{chenzu,jbiao,s.chen,dqzhang}@nuaa.edu.cn

Abstract. Recently, multi-task feature selection methods have been applied to jointly identify the disease-related brain regions for fusing information from multiple modalities of neuroimaging data. However, most of those approaches ignore the complementary label information across modalities. To address this issue, in this paper, we present a novel label-alignment-based multi-task feature selection method to jointly select the most discriminative features from multi-modality data. Specifically, the feature selection procedure of each modality is treated as a task and a group sparsity regularizer (i.e., $\ell_{2,1}$ norm) is adopted to ensure that only a small number of features to be selected jointly. In addition, we introduce a new regularization term to preserve label relatedness. The function of the proposed regularization term is to align paired within-class subjects from multiple modalities, i.e., to minimize their distance in corresponding low-dimensional feature space. The experimental results on the magnetic resonance imaging (MRI) and fluorodeoxyglucose positron emission tomography (FDG-PET) data of Alzheimer's Disease Neuroimaging Initiative (ADNI) dataset demonstrate that our proposed method can achieve better performances over state-of-the-art methods on multimodal classification of Alzheimer's disease (AD) and mild cognitive impairment (MCI).

Keywords: Alzheimer's disease · Mild cognitive impairment · Label alignment · Multi-task learning · Multi-modality

1 Introduction

Alzheimer's disease (AD) is the most common form of dementia in people over 65 years of age. It is reported that there are 26.6 million AD sufferers worldwide, and 1 in 85 people will be affected by 2050 [1]. Thus, effective and accurate diagnosis of AD and its prodromal stage (i.e., mild cognitive impairment, MCI), is very important for possible delay and early treatment of the brain disease. Lots of efforts have been made for possible identification of such changes at the early stage by leveraging neuroimaging data [3,13]. At present, several modalities of biomarkers have been proved to be sensitive to AD and MCI, such as the

© Springer International Publishing AG 2016
I. Rish et al. (Eds.): MLINI 2014, LNAI 9444, pp. 51–59, 2016.
DOI: 10.1007/978-3-319-45174-9_6

brain atrophy measured in magnetic resonance imaging (MRI) [7] and the cerebral metabolic rates of glucose measured in fluorodeoxyglucose positron emission tomography (FDG-PET) [8].

As multiple features are extracted from different imaging modalities, there may exist some irrelevant or redundant features. So, feature selection, which can be considered as the biomarker identification for AD and MCI, is commonly used to remove these redundant and irrelevant features. Some feature selection methods based on multi-modality data have been proposed for jointly selecting the most discriminative features relevant to disease. For example, Zhang et al. [12] proposed a multi-modal multi-task learning for joint feature selection for AD classification and regression. Liu et al. [5] proposed inter-modality relationship constrained multi-task feature selection for AD/MCI classification. Jie et al. [4] presented a manifold regularized multi-task feature selection method for classification of AD, and achieved the state-of-the-art performance on Alzheimer's Disease Neuroimaging Initiative (ADNI) database. However, those methods ignore the label information of data from multiple modalities, i.e., the subjects from the same class across multiple modalities should be closer in the low-dimensional feature space.

In this paper, to address this issue, we propose a novel label-alignment-based multi-task feature selection method that considers the intrinsic label relatedness among multi-modality data and preserves the complementary information conveyed by different modalities. We formulate the classification of multi-modality data as a multi-task learning (MTL) problem, where each task focuses on the classification of each modality. Specifically, two regularization items are included in the proposed model. The first item is a group Lasso regularizer [11], which ensures only a small number of features to be jointly selected across different tasks. The second item is a label-alignment regularization term, which can minimize the distance of within-class subjects from multiple modalities after projection to low-dimensional feature space leading to the selection of more discriminative features. Then, we use a multi-kernel support vector machine to fuse the above-selected features from each individual modality. The proposed method has been evaluated on ADNI dataset and obtained promising results.

The rest of this paper is organized as follows. In Sect. 2, we present the proposed label-alignment-based multi-task feature selection method in detail. Experimental results on ADNI dataset using MRI and FDG-PET biomarkers are given in Sect. 3. Finally, Sect. 4 concludes this paper and indicates points for future work.

2   Methods

2.1   Label-Alignment-Based Multi-Task Feature Selection

In this paper, we treat feature selection as a multi-task regression problem that incorporates the relationship between different modalities. Suppose we have M supervised learning tasks (i.e., the number of modalities). Denote $\boldsymbol{X}^m = [\boldsymbol{x}_1^m, \boldsymbol{x}_2^m, \ldots, \boldsymbol{x}_N^m]^T \in \mathbb{R}^{N \times d}$ as a $N \times d$ matrix that represents d

features of N training samples on the m-th task (i.e., m-th modality), and $\boldsymbol{Y} = [y_1, y_2, \ldots, y_N]^T \in \mathbb{R}^N$ as the response vector from these training subjects, where $\boldsymbol{x}_i^m$ represents feature vector of the i-th subjects of the m-th modality, and y_i is the corresponding class label (i.e., patient or normal control). Suppose $\boldsymbol{w}^m \in \mathbb{R}^d$ is the regression coefficient vector of the m-th task. Then the multi-task feature selection (MTFS) model is to solve the following objective function:

$$\min_{\boldsymbol{W}} \sum_{m=1}^{M} \|\boldsymbol{Y} - \boldsymbol{X}^m \boldsymbol{w}^m\|_2^2 + \lambda_1 \|\boldsymbol{W}\|_{2,1} \tag{1}$$

where $\boldsymbol{W} = [\boldsymbol{w}^1, \boldsymbol{w}^2, \ldots, \boldsymbol{w}^M] \in \mathbb{R}^{d \times M}$ is the weight matrix whose row $\boldsymbol{w}_j$ is the vector of coefficients associated with the j-th feature across different tasks. $\|\boldsymbol{W}\|_{2,1}$ is the $\ell_{2,1}$-norm of matrix $\boldsymbol{W}$ defined as $\|\boldsymbol{W}\|_{2,1} = \sum_{j=1}^{d} \|\boldsymbol{w}_j\|_2$ which is the sum of the ℓ_2-norms of the rows of matrix $\boldsymbol{W}$ [11]. The first term of Eq. (1) measures the empirical error on the training data while the $\ell_{2,1}$-norm encourages matrix with many zero rows. So the $\ell_{2,1}$-norm combines multiple tasks and ensures that a small number of common features will be selected across different tasks. λ_1 is a regularization parameter which balances the relative contributions of the two terms.

The MTFS model using a linear mapping function transforms the data from the original high-dimensional space to one-dimensional space. The limitation of the model is that only the relationship between data and class label for each task is considered, while the mutual dependence among data and the complementary information conveyed by different modalities are ignored, which may result in large deviations even for very similar data after mapping. To address this problem, we introduce a new regularization term called label-alignment regularization term which minimizes the distance between feature vectors of multiple modalities of the within-class subjects after feature projection:

$$\Omega = \sum_{i,j}^{N} \sum_{p,q(p<q)}^{M} \|(\boldsymbol{w}^p)^T \boldsymbol{x}_i^p - (\boldsymbol{w}^q)^T \boldsymbol{x}_j^q\|_2^2 S_{ij} \tag{2}$$

where $\boldsymbol{x}_i^p$ and $\boldsymbol{x}_j^q$ are the feature vectors of the i-th and the j-th subjects in the p-th and q-th modalities respectively. S_{ij} denotes the element of the similarity matrix $\boldsymbol{S}$ across different subjects. Here, the similarity matrix can be defined as:

$$S_{ij} = \begin{cases} 1, \text{if } \boldsymbol{x}_i^p \text{ and } \boldsymbol{x}_j^q \text{ are from the same class} \\ 0, \text{otherwise.} \end{cases} \tag{3}$$

The regularization term Eq. (2) can be explained as follows. $\|(\boldsymbol{w}^p)^T \boldsymbol{x}_i^p - (\boldsymbol{w}^q)^T \boldsymbol{x}_j^q\|_2^2 S_{ij}$ measures the distance between $\boldsymbol{x}_i^p$ and $\boldsymbol{x}_j^q$ in the projected space. It implies that, if $\boldsymbol{x}_i^p$ and $\boldsymbol{x}_j^q$ are from the same class, the distance between them should be as small as possible. Otherwise, the distance between them should be as large as possible. When $p = q$, the local geometric structrue of the same modality data are preserved during the mapping and when $p < q$, the complementary information provided from different modalities are preserved after projection of feature vectors onto the one-dimensional feature space. By incorporating the regularizer Eq. (2) into Eq. (1), we can obtain the objective function of our label-alignment-based multi-task feature selection model:

$$
\min_{\boldsymbol{W}} \quad \sum_{m=1}^{M} \| \boldsymbol{Y} - \boldsymbol{X}^m \boldsymbol{w}^m \|_2^2 + \lambda_1 \| \boldsymbol{W} \|_{2,1} + \lambda_2 \sum_{i,j}^{N} \sum_{p,q(p \leq q)}^{M} \| (\boldsymbol{w}^p)^T \boldsymbol{x}_i^p - (\boldsymbol{w}^q)^T \boldsymbol{x}_j^q \|_2^2 S_{ij}
$$

$$(4)$$

where λ_1 and λ_2 are the two positive constants that control the sparseness and the degree of preserving the distance between subjects, respectively. To optimize the problem in Eq. (4), we use Accelerated Proximal Gradient (APG) method [6] and only those features with non-zero regression coefficients are used for final classification.

2.2 Multi-modality Data Fusion and Classification

In this paper, we adopt a multi-kernel based support vector machine (SVM) method to integrate features from different modalities for classification [13]. Specifically, we calculate the linear kernels based on the features selected by the above-proposed feature selection method by using multi-modal biomarkers. Then, a combined kernel matrix is constructed by linearly combining kernels from different modalities and used in multi-kernel based SVM. The optimal parameters used for combining different kernels are determined by using a coarse-grid search through cross-validation on the training samples.

We conduct standard 10-fold cross-validation to evaluate classification performance. For each of the 10 trials, within the training data, an internal 10-fold cross-validation is performed to fine tune the parameters, i.e., the regularization parameters λ_1, λ_2 and the kernel combination parameter. The model that reaches the best performance during the inner cross-validation stage is considered as the optimal model and is adopted to classify unseen testing samples. This process is repeated 10 times independently to avoid any bias introduced by randomly partitioning dataset in the 10-fold cross-validation and the average results are reported.

Figure 1 gives a schematic illustration of our multimodal data fusion and classification pipeline, where two modalities of data (e.g., MRI and FDG-PET) are used for jointly selection features corresponding to different tasks.

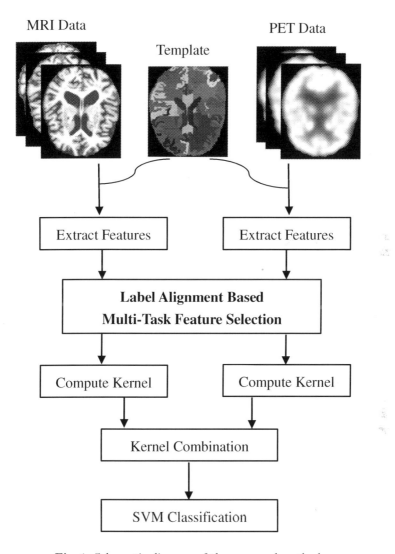

Fig. 1. Schematic diagram of the proposed method

3 Experiments

In this section, we perform a series of experiments on the Alzheimer's Disease
Neuroimaging Initiative (ADNI) database (http://adni.loni.usc.edu) to evaluate
the effectiveness of the proposed method.

3.1 Subjects and Settings

We use a total of 202 subjects with corresponding baseline MRI and FDG-PET
data from ADNI dataset: 51 AD patients, 99 MCI patients (including 43 MCI con-

56   C. Zu et al.

verters who had converted to AD within 18 months and 56 MCI non-converters), and 52 normal controls (NC). Standard image pre-processing is carried out for all MRI and FDG-PET images, including spatial distortion, skull-stripping, removal of cerebellum. Then for structural MR images, we partition each subject image into 93 manually labeled regions-of-interest (ROIs) [9] with atlas wraping. The gray matter tissue volume of these 93 ROIs is used as features extracted by the FSL package [14]. FDG-PET image of each subject is aligned onto its corresponding MR image using a grid transformation and the average intensity of each ROI in the FDG-PET image is calculated as features. Therefore, we can finally acquire 93 features from MRI image and other 93 features from PET image.

In our experiments, we compare our proposed method with multi-modality multi-kernel method [13] without feature selection (denoted as Baseline), single-modality feature selection with Lasso integrated with multi-modality multi-kernel SVM (denoted as SMFS) and multi-task feature selection method [12] (denoted as MTFS). In addition, we also concatenate all features from MRI and FDG-PET into one feature vector and perform Lasso-based feature selection and then use the standard SVM with linear kernel for classification (denoted as CONFS). For each comparison, different methods are evaluated on multiple binary classification tasks, i.e., AD vs. NC, MCI vs. NC and MCI converters (MCI-C) vs. MCI non-converters (MCI-NC), respectively. To evaluate the performances of different methods, we use four performance measures, including classification accuracy, area under receiver operating characteristic (ROC) curve (AUC), sensitivity (i.e., the proportion of patients that are correctly predicted), and specificity (i.e., the proportion of normal controls that are correctly predicted).

3.2 Results

Table 1 shows the experimental results achieved by five different methods. As can be seen from Table 1, the proposed feature selection method is always superior to other methods on three classification tasks. Specifically, our method obtains the classification accuracy of 95.51%, 82.15% and 70.50% for AD vs. NC, MCI vs. NC and MCI-C vs. MCI-NC, respectively. On the other hand, the best classification accuracy of other methods are 92.25 %, 74.34 % and 61.67 % on three tasks, respectively. Besides, we perform the standard paired t-test on the accuracies of our proposed and those of compared methods. It is shown that our

Table 1. Classification performance of all comparison methods

Method	AD vs. NC				MCI vs. NC				MCI-C vs. MCI-NC			
	ACC (%)	SEN (%)	SPE (%)	AUC	ACC (%)	SEN (%)	SPE (%)	AUC	ACC (%)	SEN (%)	SPE (%)	AUC
Baseline	91.65	92.94	90.19	0.9615	74.34	85.35	53.46	0.7764	59.67	46.28	69.64	0.6010
CONFS	91.02	90.39	91.35	0.9486	73.44	76.46	67.12	0.7802	58.44	52.33	63.04	0.6019
SMFS	92.25	92.16	92.12	0.9674	73.84	77.27	66.92	0.7745	61.67	54.19	66.96	0.6139
MTFS	92.07	91.76	92.12	0.9557	74.17	81.31	60.19	0.7758	61.61	57.21	65.36	0.6179
Proposed	**95.51**	**97.06**	**93.85**	**0.9688**	**82.15**	**86.36**	**73.85**	**0.8317**	**70.50**	**66.98**	**72.50**	**0.6857**

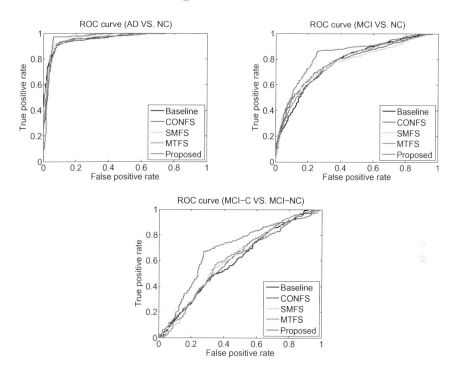

Fig. 2. ROC curves of different methods for classifications

proposed method is significantly better than the comparison methods with p values smaller than 0.05. In addition, Fig. 2 further plots the corresponding ROC curves of different methods for three classification tasks. These results demonstrate that considering the complementary label information of multi-modality data can significantly improve the classification performance, with comparison to traditional methods.

Furthermore, in Table 2, we compare our proposed method with several recent start-of-the-art methods for multimodal AD/MCI classification. Gray et al. got the classification accuracy of 89.0 %, 74.6 % and 58.0 % for AD vs. NC, MCI vs.

Table 2. Comparison on performance of different multi-modality classification methods

Method	Modalities	AD vs. NC	MCI vs. NC	MCI-C vs. MCI-NC
Gray et al. [2]	MRI+PET+ CSF+genetic	89.0 %	74.6 %	58.0 %
Westman et al. [10]	MRI+PET	91.8 %	77.6 %	68.5 %
Liu et al. [5]	MRI+PET	94.4 %	78.8 %	-
Jie et al. [4]	MRI+PET	95.0 %	79.3 %	68.9 %
Proposed	MRI+PET	**95.5 %**	**82.2 %**	**70.5 %**

NC, and MCI-C vs. MCI-NC, respectively with four different modalities (MRI + PET + CSF + genetic) [2]. When using two modalities of features (MRI + PET), Jie et al. [4] achieved the accuracy of 95.0 %, 79.3 % and 68.9 % for classification of AD vs. NC, MCI vs. NC and MCI-C vs. MCI-NC, respectively, which are inferior to our method. These results further validate the efficacy of our proposed method for multimodal AD/MCI classification.

4   Discussion

This paper addresses the problem of integrating the complementary label information to build the multi-task feature selection method for jointly selecting features from multi-modality neuroimaging data to improve AD/MCI classification. Specifically, we formulate the multi-modality classification as a multi-task learning framework and introduce the label-alignment regularization term to seek the optimal features which preserve the discriminative information between within-class subjects across multiple modalities. Experimental results demonstrate that our proposed method can achieve better performance than all conventional methods. In future work, we will extend our method to include more modalities and test other classifiers for further improvement of classification performance.

Acknowledegment. This work is supported in part by National Natural Science Foundation of China (Nos. 61422204, 61473149, 61170151), Jiangsu Natural Science Foundation for Distinguished Young Scholar (No. BK20130034), NUAA Fundamental Research Funds (No. NE2013105), the Jiangsu Qinglan Project, Natural Science Foundation of Anhui Province (No. 1508085MF125), the Open Projects Program of National Laboratory of Pattern Recognition (No. 201407361).

References

1. Brookmeyer, R., Johnson, E., Ziegler-Graham, K., Arrighi, H.M.: Forecasting the global burden of alzheimer's disease. Alzheimer's Dementia **3**(3), 186–191 (2007)
2. Gray, K.R., Aljabar, P., Heckemann, R.A., Hammers, A., Rueckert, D.: Random forest-based similarity measures for multi-modal classification of alzheimer's disease. NeuroImage **65**, 167–175 (2013)
3. Huang, S., Li, J., Ye, J., Wu, T., Chen, K., Fleisher, A., Reiman, E.: Identifying alzheimer's disease-related brain regions from multi-modality neuroimaging data using sparse composite linear discrimination analysis. In: Advances in Neural Information Processing Systems, pp. 1431–1439 (2011)
4. Jie, B., Zhang, D., Cheng, B., Shen, D.: Manifold regularized multi-task feature selection for multi-modality classification in alzheimer's disease. In: Mori, K., Sakuma, I., Sato, Y., Barillot, C., Navab, N. (eds.) MICCAI 2013, Part I. LNCS, vol. 8149, pp. 275–283. Springer, Heidelberg (2013)
5. Liu, F., Wee, C.-Y., Chen, H., Shen, D.: Inter-modality relationship constrained multi-task feature selection for AD/MCI classification. In: Mori, K., Sakuma, I., Sato, Y., Barillot, C., Navab, N. (eds.) MICCAI 2013, Part I. LNCS, vol. 8149, pp. 308–315. Springer, Heidelberg (2013)

6. Liu, J., Ye, J.: Efficient L1/Lq norm regularization. Technical report, Arizona State University (2009)
7. McEvoy, L.K., Fennema-Notestine, C., Roddey, J.C., Hagler Jr., D.J., Holland, D., Karow, D.S., Pung, C.J., Brewer, J.B., Dale, A.M.: Alzheimer disease: quantitative structural neuroimaging for detection and prediction of clinical and structural changes in mild cognitive impairment. Radiology **251**(1), 195 (2009)
8. Mosconi, L., Berti, V., Glodzik, L., Pupi, A., De Santi, S., de Leon, M.J.: Preclinical detection of alzheimer's disease using FDG-PET, with or without amyloid imaging. J. Alzheimers Dis. **20**(3), 843–854 (2010)
9. Shen, D., Davatzikos, C.: Hammer: hierarchical attribute matching mechanism for elastic registration. IEEE Trans. Med. Imaging **21**(11), 1421–1439 (2002)
10. Westman, E., Muehlboeck, J., Simmons, A., et al.: Combining MRI and CSF measures for classification of alzheimer's disease and prediction of mild cognitive impairment conversion. Neuroimage **62**(1), 229–238 (2012)
11. Yuan, M., Lin, Y.: Model selection and estimation in regression with grouped variables. J. Roy. Stat. Soc.: Ser. B (Stat. Methodol.) **68**(1), 49–67 (2006)
12. Zhang, D., Shen, D.: Multi-modal multi-task learning for joint prediction of multiple regression and classification variables in alzheimer's disease. Neuroimage **59**(2), 895–907 (2012)
13. Zhang, D., Wang, Y., Zhou, L., Yuan, H., Shen, D.: Multimodal classification of alzheimer's disease and mild cognitive impairment. Neuroimage **55**(3), 856–867 (2011)
14. Zhang, Y., Brady, M., Smith, S.: Segmentation of brain MR images through a hidden markov random field model and the expectation-maximization algorithm. IEEE Trans. Med. Imaging **20**(1), 45–57 (2001)

Leveraging Clinical Data to Enhance Localization of Brain Atrophy

João M. Monteiro[1]([⊠]), Anil Rao[1], John Ashburner[2], John Shawe-Taylor[1],
and Janaina Mourão-Miranda[1]

[1] Department of Computer Science, University College London, London, UK
{joao.monteiro,a.rao,j.shawe-taylor,j.mourao-miranda}@ucl.ac.uk
[2] Wellcome Trust Centre for Neuroimaging, Institute of Neurology,
University College London, London, UK
j.ashburner@ucl.ac.uk

Abstract. Sparse Canonical Correlation Analysis (SCCA) has been proposed to find pairs of sparse weight vectors that maximize correlations between sets of paired variables. This is done by computing one weight vector pair, deflating the correlation matrix between the views, and then repeating the process to compute the next pair. However, the deflation step used does not guarantee the orthogonality of the vector pairs. This is a very important requirement if one wishes to study the space spanned by these vectors, which should have very promising neuroscience applications. In the present work, we propose a new method for performing the deflation step in SCCA models. The ability of these vector pairs to generalize to new data was tested using an open-access dementia dataset which included T1-weighted MRI images and clinical information. The proposed method provided weight vector pairs that were both orthogonal and able to generalize to new data.

Keywords: Sparse Canonical Correlation Analysis · Matrix deflation · Neuroimaging · Dementia

1 Introduction

Pattern recognition algorithms have been successfully applied to analyse neuroimaging data for a variety of applications, including the study of neurological and psychiatric diseases. However, so far, most of these studies have focused on supervised binary classification problems, i.e. they summarize the clinical assessment into a single measure (e.g. diagnostic classification) and the output of the models is limited to a probability value and, in most cases, a binary decision (e.g. healthy/patient) [4,6,8–10,12].

Unsupervised learning approaches, such as Canonical Correlation Analysis (CCA), may provide useful insights into brain mechanisms by finding relationships between different sets of measurements (i.e. views) from the same subjects. CCA identifies a projection or latent space containing the relevant information

© Springer International Publishing AG 2016
I. Rish et al. (Eds.): MLINI 2014, LNAI 9444, pp. 60–68, 2016.
DOI: 10.1007/978-3-319-45174-9_7

in both views [13]. By studying this latent space, one can extract information regarding the underlying relationship between them.

The output of CCA consists of two sets of weight vectors that maximize the correlations between the two views. However, due to the high dimensionality of neuroimaging data, the interpretability of the image weight vectors is lost. Sparsity is imposed on the models to overcome this issue, usually by using a variation of CCA called Sparse CCA (SCCA).

Witten et al. proposed a SCCA approach with both L_1 and L_2 constraints imposed on the optimization problem [16]. The L_1 constraint favors sparse solutions, while the L_2 constraint allows for correlated features to be included in the model. This type of algorithm has gained some popularity in neuroimaging, being applied in a few studies, either directly or with some variant [2,3,14,17]. Similar approaches were also proposed slightly before the paper by Witten et al. (e.g. [11]). The SCCA algorithm acquires one pair of weight vectors by converging in an iterative process, after which it deflates the correlation matrix between the two views to acquire the next pair. However, upon performing this deflation step, we were unable to acquire a pair of weight vectors orthogonal to the first pair.

In this paper, we propose an alternative method for performing the matrix deflation step that ensures the orthogonality of the weight vectors. This should be of utmost importance if one wants to use these weight vectors to characterize the space in which the views of the data are maximally correlated. In order to test the proposed method, we implemented the SCCA algorithm described in [16] and applied it to an open-access dementia dataset consisting of T1-weighted MRI images and clinical/demographic information [7]. The weight vectors were then obtained with the deflation step described in [16] and the deflation step proposed in this paper.

2   Methods

2.1   Sparse Canonical Correlation Analysis (SCCA)

Each view was organized in a matrix where each row corresponded to a subject and each column to a feature. This was done for both image ($\boldsymbol{X}$) and clinical ($\boldsymbol{Y}$) views.

SCCA allows one to find a subspace that maximally correlates both views of a paired dataset. This is done by solving the following optimization problem [16]:

$$\max_{u,v} = \boldsymbol{u}^T \boldsymbol{X}^T \boldsymbol{Y} \boldsymbol{v}$$

$$\text{subject to} \tag{1}$$

$$\|\boldsymbol{u}\|_2^2 \leqslant 1, \ \|\boldsymbol{v}\|_2^2 \leqslant 1, \ \|\boldsymbol{u}\|_1 \leqslant c_{\mathrm{x}}, \ \|\boldsymbol{v}\|_1 \leqslant c_{\mathrm{y}}$$

where $\boldsymbol{X}^T$ and $\boldsymbol{u}^T$ denote the transposed of $\boldsymbol{X}$ and $\boldsymbol{u}$, respectively; $\boldsymbol{u}$ and $\boldsymbol{v}$ correspond to the weight vectors for the image and clinical views, respectively; and c_{x} and c_{y} correspond to the L_1 constraints on the image and clinical views, respectively. The lower the values of c_x and c_y, the stronger the L_1 constraint on the corresponding view is. Their values should be $1 \leqslant c_x \leqslant \sqrt{d_x}$ and $1 \leqslant c_y \leqslant \sqrt{d_y}$, where d_x and d_y are equal to the number of variables in X and Y, respectively.

The algorithm used to solve the optimization problem (1) is described in [16] as the following:

1. $C = X^T Y$;
2. Apply SVD to obtain the 1st estimate of v;
3. Initialize v to have L_2-norm $= 1$;
4. Iterate until convergence:
   (a) $u \leftarrow \frac{S(Cv, \Delta_x)}{\|S(Cv, \Delta_x)\|_2}$, where $\Delta_x = 0$ if it results in $|u|_1 \leqslant c_x$; otherwise, Δ_x is chosen to be a positive constraint such that $|u|_1 = c_y$;
   (b) $v \leftarrow \frac{S(C^T u, \Delta_y)}{\|S(C^T u, \Delta_y)\|_2}$, where $\Delta_y = 0$ if it results in $|v|_1 \leqslant c_y$; otherwise, Δ_y is chosen to be a positive constraint such that $|v|_1 = c_y$;
5. $d \leftarrow u^T C v$
6. $C \leftarrow C - d u v^T$

where S is the soft threshold operator defined as $S(a, c) = \text{sgn}(a)(|a| - c)_+$, where $c > 0$ is a constant and x_+ is equal to x if $x > 0$ and $x = 0$ if $x \leqslant 0$ [16].

After performing the loop described in step 4, the weight vectors are obtained. If one wants to obtain additional sparse weight vectors, then matrix C has to be deflated using the last two steps. However, we propose an alternative way of doing this.

If we consider $\widetilde{C}$ to be the matrix whose variance is not explained by u and v (i.e. the one we want to obtain, so that we can calculate the new weight vector pair), then:

$$\widetilde{C} = \widetilde{X}^T \widetilde{Y} \tag{2}$$

Note that if the weight vectors obtained are orthogonal, then: $\widetilde{X} u = 0$ and $\widetilde{Y} v = 0$.

$\widetilde{X}$ and $\widetilde{Y}$ can be re-written as:

$$\widetilde{X} = X(I - u u^T) \text{ and } \widetilde{Y} = Y(I - v v^T) \tag{3}$$

By substituting (3) in (2) and doing some manipulations, one arrives at:

$$\widetilde{C} = C - C v v^T - u u^T C + d u v^T \tag{4}$$

Step 6 in the algorithm can then be substituted by: $C \leftarrow \widetilde{C}$.

The first three pairs of weight vectors obtained were used to project test data in order to evaluate the model's performance, i.e. ability to learn associations between the two views that can be generalized to new data:

$$P_x = X_{\text{test}} U \quad \text{and} \quad P_y = Y_{\text{test}} V \tag{5}$$

where P_x and P_y denote the projection of the image and clinical views, respectively; U and V are the column vector matrices containing the three weight vectors. In order to calculate a summary of the correlation between the three components of the projections simultaneously, the Hilbert-Schmidt Independence Criterion (HSIC) [5] was used.

2.2 Validation Framework

A validation framework was used to test the SCCA model for its ability to generalize to new data. This was done by projecting the two views of the test data onto the SCCA weight vectors computed using the training data (5) and computing the correlation between the projected views. The procedure used to validate the model was the following:

1. Create 100 random splits of the data: 50 % for training and 50 % for testing;
2. For one split, run SCCA on the training data, and project both views of the test data using the SCCA weight vectors for each view (5). Then, calculate the correlation between the projected views using HSIC;
3. Randomly permute the order of the rows in the clinical test data matrix, project both views of the test data using the SCCA weight vectors for each view (5). Then, calculate the correlation between the projected views using HSIC;
4. Go back to step 3 and repeat 10000 times;
5. Check how many times the absolute values of the correlation calculated in step 3 (permuted SCCA correlation) were greater or equal to the one calculated in step 2 (original SCCA correlation). If the fraction is lower than 0.05, then, the correlation was considered as being statistically significant.
6. Go back to step 2 and select the next split;

In order to summarize the results of the validation framework, the percentage of splits that were considered statistically significant (step 5) was determined. Note that by using this framework (i.e. permuting the test data), we are testing the significance of the learnt weight vectors (their performance on test data), rather than the ability of SCCA to learn weight vectors. The latter would require the permutations to be performed on the whole dataset, before the SCCA weight vectors are obtained.

2.3 Dataset

A subset from the "OASIS: Longitudinal MRI Data in Nondemented and Demented Older Adults" dataset (www.oasis-brains.org) was used. The dataset includes T1-weighted MRI scans from subjects at different time points, with the corresponding clinical information [7]. In this study, a subset of 142 subjects from the first time point was used. These included 58 male subjects and 84 female subjects with an average age of 75.4 years $\pm$ 7.7 years. Among these, 72 were considered as being healthy, 56 as demented, and 14 as "converted". The latter ones were considered as healthy on the first time point, but latter developed dementia.

All images were preprocessed using SPM12b [15]. The first step was to average all the repeats for each session followed by a grey matter segmentation, then, the segmented images were registered using DARTEL [1], normalized to MNI space with a voxel size of [2 2 2] and smoothed with a Gaussian kernel with a full width at half maximum (FWHM) of [8 8 8]. In addition, the head size was regressed out of the data and a mask was applied to select voxels that had a probability of being grey matter equal or above 20 %. The grey matter probability maps within the mask were used as the first view of SCCA.

For each subject, the following clinical information was used as the second view of SCCA: age, social economic status, education, Mini-Mental State Examination (MMSE) and Clinical Dementia Rating (CDR). All features in both views were mean-centered and normalized.

2.4   Previous Deflation Procedure vs. Proposed Procedure

To see the difference between the previous and the proposed deflation procedures (4), each SCCA procedure was applied to compute three weight vector pairs. Then, we checked if these weight vectors were orthogonal and used them to project the data matrices from each view onto them, i.e. the image data matrix was projected onto the three image weight vectors and the clinical data matrix was projected onto the three clinical weight vectors. The resulting projections were then plotted for each deflation procedure.

Based on previous results, the regularization parameters were set to $c_x = 50$ and $c_y = \sqrt{5}$ for both deflation procedures, since they provided a statistically significant solution with an adequate amount of selected voxels for visualizing the effects of the different procedures. Due to space constraints, we will not be able to present these results. However, it should be noted that finding the optimal level of sparsity is not the objective of this paper, the main focus lies on the effects of the weight vector orthogonalization, which are observed using different levels of sparsity.

3   Results and Discussion

The weight vector groups obtained using the previous deflation procedure were not orthogonal. When calculating the inner products between the vectors, the range of the absolute values obtained was [0.24, 1], with a median of 0.82. In contrast, the weight vectors obtained with the proposed deflation step were orthogonal, all the inner products were approximately 0. Furthermore, the inner products calculated for each split of the validation procedure described in Sect. 2.2 followed the same tendency. The range of the absolute values of the inner products using the previous deflation procedure was [0.01, 1.00] with a median of 0.89, while the range and median for the proposed deflation procedure were [0.00, 0.05] and 0.00, respectively.

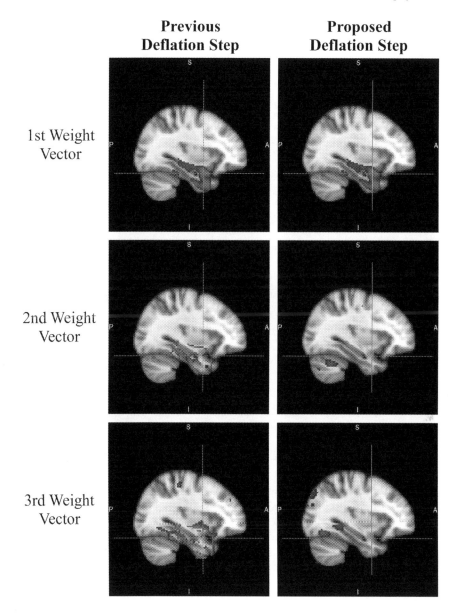

Fig. 1. First three image weight vectors calculated using each deflation step. Red regions correspond to positive weight values and blue regions correspond to negative weight values.

Figure 1 shows the weight maps for the first three weight vectors calculated using each deflation step. The second weight vector calculated using the previous deflation step selects regions surrounding the ones selected by the previous weight vector, and the third weight vector selects regions surrounding regions selected by the second weight vector. This suggests that SCCA is detecting the same

a)

b)

c)

d)

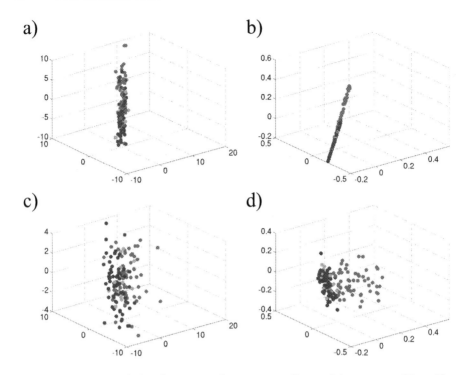

Fig. 2. Projections of the data onto the corresponding weight vectors. Blue: Non-Demented; Red: Demented; Green: "Converted". (**a**) Previous deflation - projections using the image weights; (**b**) previous deflation - projections using the clinical weights; (**c**) proposed deflation - projections using the image weights; (**d**) proposed deflation - projections using the clinical weights. (Color figure online)

effect in the data for each weight vector pair. By enforcing the orthogonality of the weight vectors, the effects described by the previous weight vectors are removed from the data, allowing new effects to be detected.

The nonorthogonality of the weight vectors leads to projections that can be seen in Fig. 2(a) and (b), where the projections are aligned along a line. However, this is not the case when the proposed procedure is used (Fig. 2(c) and (d)). In this case, the weight vectors obtained were orthogonal, which leads to projections that are able to capture the variance of the data.

The models were both tested for significance using the procedure described in Sect. 2.2 and they were both significant ($p < 0.05$) in 100 % of the splits. However, this only means that the vectors could not have been obtained by chance alone. The group of paired weight vectors obtained using the previous deflation might be significant, but the vectors are still not orthogonal (which should not be the case when matrix deflation is applied).

4 Conclusion

The results showed that the proposed method is able to calculate statistically significant orthogonal weight vectors. By guaranteeing the orthogonality of the image weight vectors, new brain regions that correlate with the clinical variables could be found. Therefore, we believe that using this method is preferable when applying SCCA with matrix deflation to study neuroimaging data.

In order to capture the variance of the data in the subspace defined by the weight vectors, it is very important to guarantee the orthogonality of these vectors. The analysis of the space spanned by the weight vectors should provide very useful information regarding the underlying relationship between the two views. Future work should include the study of this space in order to find any information that might improve the understanding of brain function or assist with the diagnosis of diseases related with the brain.

Acknowledgments. João M. Monteiro was supported by a PhD scholarship awarded by Fundação para a Ciência e a Tecnologia (SFRH/BD/88345/2012).

Janaina Mourão-Miranda was supported by the Wellcome Trust under grants No. WT086565/Z/08/Z and No. WT102845/Z/13/Z.

The authors would like to acknowledge the OASIS dataset, which was funded by the following grants: P50 AG05681, P01 AG03991, R01 AG021910, P50 MH071616, U24 RR021382, R01 MH56584.

References

1. Ashburner, J.: A fast diffeomorphic image registration algorithm. Neuroimage **38**(1), 95–113 (2007)
2. Boutte, D., Liu, J.: Sparse canonical correlation analysis applied to fMRI and genetic data fusion. In: 2010 IEEE International Conference on Bioinformatics and Biomedicine (BIBM), pp. 422–426. IEEE (2010)
3. Chi, E.C., Allen, G.I., Zhou, H., Kohannim, O., Lange, K., Thompson, P.M.: Imaging genetics via sparse canonical correlation analysis. In: 2013 IEEE 10th International Symposium on Biomedical Imaging (ISBI), pp. 740–743. IEEE (2013)
4. Ecker, C., Marquand, A., Mourão-Miranda, J., Johnston, P., Daly, E.M., Brammer, M.J., Maltezos, S., Murphy, C.M., Robertson, D., Williams, S.C., et al.: Describing the brain in autism in five dimensions–magnetic resonance imaging-assisted diagnosis of autism spectrum disorder using a multiparameter classification approach. J. Neurosci. **30**(32), 10612–10623 (2010)
5. Gretton, A., Bousquet, O., Smola, A.J., Schölkopf, B.: Measuring statistical dependence with Hilbert-Schmidt norms. In: Jain, S., Simon, H.U., Tomita, E. (eds.) ALT 2005. LNCS (LNAI), vol. 3734, pp. 63–77. Springer, Heidelberg (2005)
6. Klöppel, S., Stonnington, C.M., Chu, C., Draganski, B., Scahill, R.I., Rohrer, J.D., Fox, N.C., Jack, C.R., Ashburner, J., Frackowiak, R.S.: Automatic classification of MR scans in Alzheimer's disease. Brain **131**(3), 681–689 (2008)
7. Marcus, D.S., Fotenos, A.F., Csernansky, J.G., Morris, J.C., Buckner, R.L.: Open access series of imaging studies: longitudinal MRI data in nondemented and demented older adults. J. Cogn. Neurosci. **22**(12), 2677–2684 (2009)

8. Mourão-Miranda, J., Bokde, A.L., Born, C., Hampel, H., Stetter, M.: Classifying brain states and determining the discriminating activation patterns: support vector machine on functional MRI data. NeuroImage **28**(4), 980–995 (2005)

9. Nouretdinov, I., Costafreda, S.G., Gammerman, A., Chervonenkis, A., Vovk, V., Vapnik, V., Fu, C.H.: Machine learning classification with confidence: application of transductive conformal predictors to MRI-based diagnostic and prognostic markers in depression. Neuroimage **56**(2), 809–813 (2011)

10. Orrù, G., Pettersson-Yeo, W., Marquand, A.F., Sartori, G., Mechelli, A.: Using support vector machine to identify imaging biomarkers of neurological and psychiatric disease: a critical review. Neurosci. Biobehav. Rev. **36**(4), 1140–1152 (2012)

11. Parkhomenko, E., Tritchler, D., Beyene, J.: Sparse canonical correlation analysis with application to genomic data integration. Stat. Appl. Genet. Mol. Biol. **8**(1), 1–34 (2009)

12. Rao, A., Lee, Y., Gass, A., Monsch, A.: Classification of Alzheimer's disease from structural MRI using sparse logistic regression with optional spatial regularization. In: 2011 Annual International Conference of the IEEE Engineering in Medicine and Biology Society, EMBC, pp. 4499–4502. IEEE (2011)

13. Shawe-Taylor, J., Cristianini, N.: Kernel Methods for Pattern Analysis. Cambridge University Press, Cambridge (2004)

14. Wan, J., et al.: Hippocampal surface mapping of genetic risk factors in AD via sparse learning models. In: Fichtinger, G., Martel, A., Peters, T. (eds.) MICCAI 2011, Part II. LNCS, vol. 6892, pp. 376–383. Springer, Heidelberg (2011)

15. Wellcome Trust Centre for Neuroimaging: SPM software - Statistical Parametric Mapping. http://www.fil.ion.ucl.ac.uk/spm/software/

16. Witten, D.M., Tibshirani, R., Hastie, T.: A penalized matrix decomposition, with applications to sparse principal components and canonical correlation analysis. Biostatistics, kxp008 (2009)

17. Witten, D.M., Tibshirani, R.J.: Extensions of sparse canonical correlation analysis with applications to genomic data. Stat. Appl. Genet. Mol. Biol. **8**(1), 1–27 (2009)

Estimating Learning Effects: A Short-Time Fourier Transform Regression Model for MEG Source Localization

Ying Yang$^{(\boxtimes)}$, Michael J. Tarr, and Robert E. Kass

Carnegie Mellon University, Pittsburgh, USA
yingyan1@andrew.cmu.edu, michaeltarr@cmu.edu, kass@stat.cmu.edu

Abstract. Magnetoencephalography (MEG) has a high temporal resolution well-suited for studying perceptual learning. However, to identify *where* learning happens in the brain, one needs to apply source localization techniques to project MEG sensor data into brain space. Previous source localization methods, such as the short-time Fourier transform (STFT) method by Gramfort et al. [6] produced intriguing results, but they were not designed to incorporate trial-by-trial learning effects. Here we modify the approach in [6] to produce an STFT-based source localization method (STFT-R) that includes an additional regression of the STFT components on covariates such as the behavioral learning curve. We also exploit a hierarchical L_{21} penalty to induce structured sparsity of STFT components and to emphasize signals from regions of interest (ROIs) that are selected according to prior knowledge. In reconstructing the ROI source signals from simulated data, STFT-R achieved smaller errors than a two-step method using the popular minimum-norm estimate (MNE), and in a real-world human learning experiment, STFT-R yielded more interpretable results about what time-frequency components of the ROI signals were correlated with learning.

1 Introduction

Magnetoencephalography (MEG) [9] has a high temporal resolution well-suited for studying the neural bases of perceptual learning. By regressing MEG signals on covariates, for example, trial-by-trial behavioral performance, we can identify how neural signals change with learning. Based on Maxwell's equations [8], MEG sensor data can be approximated by a linear transform of the underlying neural signals in a "source space", often defined as $\sim 10^4$ source points distributed on the cortical surfaces. Solving the inverse of this linear problem ("source localization") facilitates identifying the neural sites of learning. However, this inverse problem is underspecified, because the number of sensors (~ 300) is much smaller than the number of source points. Many source localization methods use an L_2 penalty for regularization at each time point (minimum-norm estimate [8], dSPM [3] and sLORETA [15]). These methods, however, may give noisy solutions in that they ignore the temporal smoothness of the MEG signals. Other methods have been proposed to capture the temporal structure (e.g. [4,13]), among which, a

© Springer International Publishing AG 2016
I. Rish et al. (Eds.): MLINI 2014, LNAI 9444, pp. 69–82, 2016.
DOI: 10.1007/978-3-319-45174-9_8

sparse short-time Fourier transform (STFT) method by Gramfort et al. [6] yields solutions that are spatially sparse and temporally smooth.

With L_2 methods such as the minimum-norm estimate (MNE), one can study learning effects in a two-step procedure: (1) obtain source time series in each trial; (2) regress some features of the time series on the covariates. However, these methods may give noisy solutions due to lack of smoothness. To address this, we might want to regress the STFT components in [6] on the covariates in a two-step procedure, but being designed for single-trial data, [6] may not provide consistent sparse structures across trials. Additionally, in cases with pre-defined regions of interest (ROIs) that are theoretically important in perceptual learning, for example, "face-selective" areas [5,12,16], it is not desirable to shrink all source points equally to zero as in MNE. Instead, it may be useful to assign weighted penalties to emphasize the ROIs.

Here we modify the model in [6] to produce a new method (STFT-R) to estimate learning effects in MEG. We represent the source signals with STFT components and assume the components have a linear relationship with the covariates. To solve the regression coefficients of STFT components, we design a hierarchical group lasso (L_{21}) penalty [11] of three levels to induce structured sparsity. The first level partitions source points based on ROIs, allowing different penalties for source points within ROIs and outside ROIs; then for each source point, the second level encourages sparsity over time and frequency on the regression coefficients of the STFT components, and finally for each STFT component, the third level induces sparsity over the coefficients for different covariates. We derive an algorithm with an active-set strategy to solve STFT-R, and compare STFT-R with an alternative two-step procedure using MNE on both simulated and human experimental data.

2   Methods

Model. Assume we have n sensors, m source points, T time points in each trial, and q trials together. Let $\boldsymbol{M}^{(r)} \in \mathbb{R}^{n \times T}$ be the sensor time series we observe in the rth trial, and $\boldsymbol{G} \in \mathbb{R}^{n \times m}$ be the known linear operator ("forward matrix") that projects source signals to sensor space. Following the notation in [6], let $\boldsymbol{\Phi}^H \in \mathbb{C}^{s \times T}$ be s pre-defined STFT dictionary functions at different frequencies and time points (see Appendix 1). Suppose we have p covariates (e.g. a behavioral learning curve, or non-parametric spline basis functions), we write them into a design matrix $\boldsymbol{X} \in \mathbb{R}^{q \times p}$, which also includes an all-one column to represent the intercept. Besides the all one column, all other columns have zero means. Let the scalar $X_k^{(r)} = \boldsymbol{X}(r,k)$ be the kth covariate in the rth trial. When we represent the time series of the ith source point with STFT, we assume each STFT component is a linear function of the p covariates: the jth STFT component in the rth trial is $\sum_{k=1}^{p} X_k^{(r)} Z_{ijk}$, where the regression coefficients Z_{ijk}'s are to be solved. We use a complex tensor $\boldsymbol{Z} \in \mathbb{C}^{m \times s \times p}$ to denote the Z_{ijk}'s, and use $\boldsymbol{Z}_k \in \mathbb{C}^{m \times s}$ to denote each layer of $\boldsymbol{Z}$. Our STFT-R model reads

$$M^{(r)} = G \left(\sum_{k=1}^{p} X_k^{(r)} Z_k \right) \Phi^H + E^{(r)} \quad \text{for} \quad r = 1, \cdots, q.$$

where the error $E^{(r)} \in \mathbb{R}^{n \times T}$ is an i.i.d random matrix for each trial. To solve for Z, we minimize the sum of squared prediction error across q trials, with a hierarchical L_{21} penalty Ω on Z:

$$\min_{Z} \left(\frac{1}{2} \sum_{r=1}^{q} \| M^{(r)} - G(\sum_{k=1}^{p} X_k^{(r)} Z_k) \Phi^H \|_F^2 + \Omega(Z, \alpha, \beta, \gamma, w) \right) \quad (1)$$

where $\| \cdot \|_F$ is the Frobenius norm and

$$\Omega(Z, \alpha, \beta, \gamma, w) = \alpha \sum_{l} w_l \sqrt{\sum_{i \in \mathcal{A}_l} \sum_{j=1}^{s} \sum_{k=1}^{p} |Z_{ijk}|^2} \quad (2)$$

$$+ \beta \sum_{i=1}^{m} \sum_{j=1}^{s} \sqrt{\sum_{k=1}^{p} |Z_{ijk}|^2} \quad (3)$$

$$+ \gamma \sum_{i=1}^{m} \sum_{j=1}^{s} \sum_{k=1}^{p} |Z_{ijk}|. \quad (4)$$

The penalty Ω involves three terms corresponding to three levels of nested groups, and α, β and γ are tuning parameters. On the first level in (2), each group under the square root either consists of coefficients for all source points within one ROI, or coefficients for one single source point outside the ROIs. Therefore we have N_α groups, denoted by $\mathcal{A}_l, l = 1, \cdots, N_\alpha$, where N_α is the number of ROIs plus the number of source points outside the ROIs. Such a structure encourages the source signals outside the ROIs to be spatially sparse and thus reduces computational cost. With a good choice of weights for the N_α groups, $w = (w_1, w_2, \ldots w_{N_\alpha})^T$, we can also make the penalty on coefficients for source points within the ROIs smaller than that on coefficients for source points outside the ROIs. On the second level, for each source point i, the term (3) groups the p regression coefficients for the jth STFT component under the square root, inducing sparsity over time points and frequencies. Finally, on the third level, (4) adds an L_1 penalty on each Z_{ijk} to encourage sparsity on the p covariates, for each STFT component of each source point.

The FISTA Algorithm. We use the fast iterative shrinkage-thresholding algorithm (FISTA [2]) to solve (1), with a constant step size, following [6]. Let z be a vector that is concatenated by all entries in Z, and let y be a vector of the same size. In each FISTA step, we need the proximal operator associated with the hierarchical penalty Ω:

$$\arg\min_{\mathbf{z}}(\frac{1}{2}\|\mathbf{z}-\mathbf{y}\|^2+\Omega(\mathbf{z},\alpha,\beta,\gamma,\mathbf{w})) = \arg\min_{\mathbf{z}}(\frac{1}{2}\|\mathbf{z}-\mathbf{y}\|^2+\sum_{h=1}^{N}\lambda_h\|\mathbf{z}|_{g_h}\|_2) \quad (5)$$

where we concatenate all of the nested groups on the three levels in Ω into an ordered list $\{g_1, g_2, \cdots, g_N\}$ and denote the penalty on group g_h by λ_h. For example, $\lambda_h = \alpha w_l$ if g_h is the lth group on the first level, $\lambda_h = \beta$ if g_h is on the second level, and $\lambda_h = \gamma$ if g_h is on in the third level. $\{g_1, g_2, \cdots, g_N\}$ is obtained by listing all the third level groups, then the second level and finally the first level, such that if h_1 is before h_2, then $g_{h_1} \subset g_{h_2}$ or $g_{h_1} \cap g_{h_2} = \emptyset$. Let $\mathbf{z}|_{g_h}$ be the elements of $\mathbf{z}$ in group g_h. As proved in [11], (5) is solved by composing the proximal operators for the L_{21} penalty on each g_h, following the order in the list; that is, initialize $\mathbf{z} \leftarrow \mathbf{y}$, for $h = 1, \cdots N$ in the ordered list,

$$\mathbf{z}|_{g_h} \leftarrow \begin{cases} \mathbf{z}|_{g_h}(1 - \lambda_h/\|\mathbf{z}|_{g_h}\|_2) & \text{if } \|\mathbf{z}|_{g_h}\|_2 > \lambda_h \\ 0 & \text{otherwise} \end{cases}$$

Algorithm 1. FISTA algorithm given the Lipschitz constant L

Data: $L, f(\mathbf{z}) = \frac{1}{2}\sum_{r=1}^{q} \|\mathbf{M}^{(r)} - \mathbf{G}\left(\sum_{k=1}^{p} X_k^{(r)} \mathbf{Z}_k\right)\boldsymbol{\Phi}^H\|_F^2, \Omega(\mathbf{z}) = \Omega(\mathbf{Z}, \alpha, \beta, \gamma, \mathbf{w})$

Result: the optimal solution $\mathbf{z}$

initialization: $\mathbf{z}_0, \zeta = 1, \zeta_0 = 1, \mathbf{y} \leftarrow \mathbf{z}_0, \mathbf{z} \leftarrow \mathbf{z}_0$;

while *change of $\mathbf{z}$ in two iterations is not small enough* **do**

    $\mathbf{z}_0 \leftarrow \mathbf{z}$; Compute $\nabla f(\mathbf{y})$;

    Apply the proximal operator

    $\mathbf{z} = \arg_{\mathbf{x}} \min(\frac{1}{2}\|\mathbf{x} - (\mathbf{y} - \frac{1}{L}\nabla f(\mathbf{y}))\|^2 + \frac{1}{L}\Omega(\mathbf{x}))$;

    $\zeta_0 \leftarrow \zeta; \zeta \leftarrow \frac{1+\sqrt{4\zeta_0^2+1}}{2}; \mathbf{y} \leftarrow \mathbf{z} + \frac{\zeta_0-1}{\zeta}(\mathbf{z} - \mathbf{z}_0)$;

end

Details of FISTA are shown in Algorithm 1, where $\mathbf{y}$ and $\mathbf{z}_0$ are auxiliary variables of the same shape as $\mathbf{z}$, and ζ, ζ_0 are constants used to accelerate convergence. The gradient of $f(\mathbf{z})$ is computed in the following way: $\frac{\partial f}{\partial \mathbf{Z}_k} = -\mathbf{G}^T \sum_{r=1}^{q} X_k^{(r)} \mathbf{M}^{(r)} \boldsymbol{\Phi} + \mathbf{G}^T \mathbf{G}(\sum_{r=1}^{q} X_k^{(r)} \sum_{k'=1}^{p} \mathbf{Z}_{k'} X_{k'}(r))\boldsymbol{\Phi}^H \boldsymbol{\Phi}$. We use the power iteration method in [6] to compute the Lipschitz constant of the gradient.

The Active-Set Strategy. In practice, it is expensive to solve the original problem in (1). Thus we derive an active-set strategy (Algorithm 2), according to Chap. 6 in [1]: starting with a union of some groups on the first level ($J = \cup_{l\in\mathcal{B}}\mathcal{A}_l, \mathcal{B} \subset \{1, \cdots, N_\alpha\}$), we compute the solution to the problem constrained on J, then examine whether it is optimal for the original problem by checking whether the Karush-Kuhn-Tucker(KKT) conditions are met, if yes, we accept it, otherwise, we greedily add more groups to J and repeat the procedure.

Let $\mathbf{z}$ denote the concatenated $\mathbf{Z}$ again, and let diagonal matrix $\mathbf{D}_h$ be a filter to select the elements of $\mathbf{z}$ in group g_h (i.e. entries of $\mathbf{D}_h\mathbf{z}$ in group g_h

are equal to $z|_{g_h}$, and entries outside g_h are 0). Given a solution z_0, the KKT conditions are

$$\nabla f(z)_{z=z_0} + \sum_h D_h \xi_h = 0, \text{ and } \begin{cases} \xi_h = \lambda_h \dfrac{D_h z_0}{\|D_h z_0\|_2} & \text{if } \|D_h z_0\|_2 > 0, \\ \|\xi_h\|_2 \leq \lambda_h & \text{if } \|D_h z_0\|_2 = 0 \end{cases}$$

where $\xi_h, h = 1, \cdots, N$ are Lagrange multipliers of the same shape as z. We defer the derivations to Appendix 2.

We minimize the following problem

$$\min_{\xi_h, \forall h} \frac{1}{2} \|\nabla f(z)_{z=z_0} + \sum_h D_h \xi_h\|_2^2,$$

$$\text{subject to } \begin{cases} \xi_h = \lambda_h \dfrac{D_h z_0}{\|D_h z_0\|_2} & \text{if } \|D_h z_0\|_2 > 0, \\ \|\xi_h\|_2 \leq \lambda_h & \text{if } \|D_h z_0\|_2 = 0 \end{cases}$$

and use $\frac{1}{2}\|\nabla f(z)_{z=z_0} + \sum_h D_h \xi_h\|_2^2$ at the optimum to measure the violation of KKT conditions. Additionally, we use the $\frac{1}{2}\|(\nabla f(z)_{z=z_0} + \sum_h D_h \xi_h)|_{\mathcal{A}_l}\|_2^2$, constrained on each non-active first-level group $\mathcal{A}_l \not\subset J$, as a measurement of violation for the group.

Algorithm 2. Active-set strategy

initialization: choose initial J and initial solution Z; compute the KKT
   violation for each $\mathcal{A}_l \not\subset J$;
while *the total KKT violation is not small enough* **do**
      add 50 non-active groups that have the largest KKT violations to J;
      compute a solution to the problem constrained on J using FISTA ;
      compute the KKT violation for each $\mathcal{A}_l \not\subset J$;
end

L_2 Regularization and Bootstrapping. The hierarchical L_{21} penalty may give biased results [6]. To reduce bias, we computed an L_2 solution constrained on the non-zero entries of the hierarchical L_{21} solution. Tuning parameters in the L_{21} and L_2 models were selected to minimize cross-validated prediction error.

To obtain the standard deviations of the regression coefficients in Z, we performed a data-splitting bootstrapping procedure. The data was split to two halves (odd and even trials). On the first half, we obtained the hierarchical L_{21} solution, and on the second half, we computed an L_2 solution constrained on the non-zero entries of the hierarchical L_{21} solution. Then we plugged in this L_2 solution Z to obtain residual sensor time series of each trial on the second half of the data ($R^{(r)} = M^{(r)} - G(\sum_{k=1}^p X_k^{(r)} Z_k) \Phi^H$). We rescaled the residuals according to the design matrix X [17]. Let $X_r = X(r,:)^T = (X_1^{(r)}, X_2^{(r)}, \cdots, X_p^{(r)})^T$, and

$h_r = X_r^T (\boldsymbol{X}^T \boldsymbol{X})^{-1} X_r$. The residual in the rth trial was rescaled by $1/(1-h_r)^{0.5}$. The re-sampled residuals $\boldsymbol{R}^{(r)*}$s were random samples with replacement from $\{\boldsymbol{R}^{(r)}/(1-h_r)^{0.5}, r = 1, \cdots, q\}$ and the bootstrapped sensor data for each trial were

$$\boldsymbol{M}^{(r)*} = \boldsymbol{G}(\sum_{k=1}^{p} X_k^{(r)} \boldsymbol{Z}_k)\boldsymbol{\Phi}^H + \boldsymbol{R}^{(r)*}$$

After B re-sampling procedures, for each bootstrapped sample, we re-estimated the solution to the L_2 problem constrained on the non-zero entries again, and the best L_2 parameter was determined by a 2-fold cross-validation.

3   Results

Simulation. On simulated data, we compared STFT-R with an alternative two-step MNE method (labelled as MNE-R), that is, (1) obtain MNE source solutions for each trial; (2) apply STFT and regress the STFT components on the covariates.

We performed simulations using "mne-python" [7], which provided a sample dataset, and a source space that consisted of 7498 source points perpendicular to the gray-white matter boundary, following the major current directions that MEG is sensitive to. Simulated source signals were constrained in four regions in the left and right primary visual and auditory cortices (Aud-lh, Aud-rh, Vis-lh and Vis-rh, Fig. 1(a)). All source points outside the four regions were set to zero. To test whether STFT-R could emphasize regions of interest, we treated Aud-lh and Aud-rh as the target ROIs and Vis-lh and Vis-rh as irrelevant signal sources. The noiseless signals were low-frequency Gabor functions (Fig. 1(b)), whose amplitude was a linear function of a sigmoid curve (simulated "behavorial learning curve", Fig. 1(c)). We added Gaussian process noise on each source point in the four regions independently for each trial. The marginal standard deviation of this noise at each time point was defined as *noise level*. We ran simulations with the noise level being 0.1, 0.3 and 0.5, where the units were 10 nA. We also simulated the sensor noise as multivariate Gaussian noise filtered by a 5th order infinite impulse response (IIR) filter. The filter and covariance matrix of the sensor noise were estimated from the sample data. We controlled the signal-to-noise ratios (SNRs) in Decibel when adding sensor noise to the simulated data in each trial[1].

We ran 5 independent simulations for $SNR \in \{0.5, 1\}$ and *noise level* $\in \{0.1, 0.3, 0.5\}$, with 20 trials (length of time series $T = 100$, sampling rate $= 100$ Hz, window size of the STFT $= 160$ ms and step size $\tau_0 = 40$ ms). With only one covariate (the sigmoid curve), we fit an intercept and a slope for each

---

[1] Note that in this case, the variance of sensor noise in each trial was proportional to the source signals. This violated the i.i.d sensor noise assumption in both STFT-R and MNE-R. We compared performance of the two methods in tolerating such heteroskedasticity.

STFT component. Before applying both methods, we pre-whitened the correlation between sensors. In STFT-R, the weights for α in the ROI groups were set to zero, and the weights in the non-ROI groups were equal and summed to 1. We tuned the penalization parameters α, β in STFT-R. For γ, because the true slope and intercept were equal in the simulation, we did not need a large γ to select between the slope and intercept, therefore we fixed γ to a small value to reduce the time for parameter tuning. The L_2 penalty parameter in MNE-R was also selected via cross-validation. We used $B = 20$ in bootstrapping.

We reconstructed the source signals in each trial using the estimated $\mathbf{Z}$. Note that true source currents that were close to each other could have opposite directions due to the folding of sulci and gyri, and with limited precision of the forward matrix, the estimated signal could have an opposite sign to the truth. Therefore we "rectified" the reconstructed signals and the true noiseless signals by taking their absolute values, and computed the mean squared error (MSE) on the absolute values. Figure 1(d) shows estimated source signals in the target

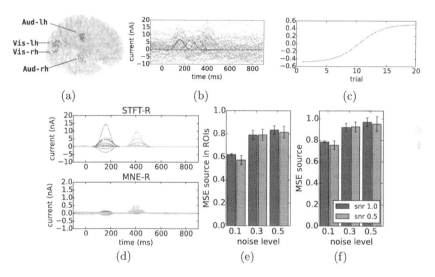

Fig. 1. Simulation results: source signal reconstruction. (a), Target ROIs: Aud-lh (red), Aud-rh (yellow) and irrelevant regions: Vis-lh (blue), Vis-rh (green). (b), The simulated source signals with Gaussian process noise in the 20th trial. Each curve represents one source point. The thicker curves show the noiseless true signals (solid: target ROIs, dashed: irrelevant regions). *noise level* = 0.5. (c), The simulated "behavioral learning curve". (d), Estimates of source signals (*noise level* = 0.5, *SNR* = 0.5) in the 20th trial by STFT-R and MNE-R, in Aud-lh (red) and Aud-rh (yellow). Each curve represents one source point. Note the scale for MNE-R is $< 1/10$ of the truth. (e) and (f), Ratios of rectified MSE (STFT-R over MNE-R) for source points within the target ROIs (e) and for all source points (f). The bars show averaged ratios across 5 independent runs of simulation, and the error bars show standard errors of the averaged ratios. (Color figure online)

ROIs (red and yellow) by the two methods in the 20th trial ($SNR = 0.5$, *noise level* = 0.5). Noticing the scales, we found that MNE-R shrank the signals much more than STFT-R. We show the ratios of the rectified MSEs of STFT-R to the rectified MSEs of MNE-R for source points within the ROIs (Fig. 1(e)), and for all source points in the source space (Fig. 1(f)). Compared with MNE-R, STFT-R reduced the MSE within the ROIs by about $20 \sim 40\%$ (Fig. 1(e)). STFT-R also reduced the MSE of all the source points by about 20% in cases with low *noise levels* (0.1) (Fig. 1(f)). The MSE reduction was larger when *noise level* was small.

To visualize which time-frequency components were correlated with the covariate, we computed the T-statistic for each slope coefficient of each STFT component, defined as the estimated coefficient divided by the bootstrapped standard error. Again, since our estimate could have an opposite sign to the true signals, we rectified the T-statistics by using their absolute values. We first

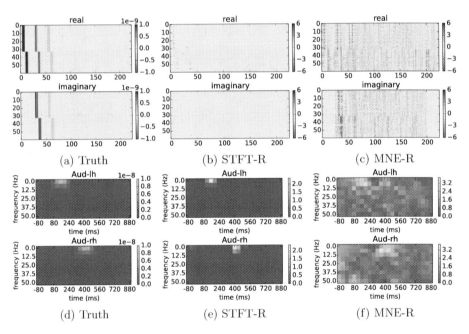

Fig. 2. Simulation results: inference of regression. $SNR = 0.5$, *noise level* = 0.5. (a), The true slope coefficients of the regression. The vertical axis corresponds to the indices of source points. Source points from the two ROIs are concatenated. The horizontal axis corresponds to the indices of frequency × time components, where 0–24 are 25 time points in 0 Hz, 25–49 in 6.25 Hz, etc. The upper and lower plots show the real and imaginary parts of the complex coefficients. (b) and (c), The T-statistics for each STFT components, by STFT-R (b) and by MNE-R(c). (d), Averaged absolute values of the real and imaginary parts of the true slope coefficients across source points in each ROI. (e) and (f), *Averaged absolute T* for each STFT component in the two ROIs by STFT-R (e) and MNE-R (f).

averaged the absolute values of the T-statistics for the real and imaginary parts of each STFT component, and then averaged them across all non-zero source points in an ROI, for each STFT component. We call these values *averaged absolute T*s.

In Fig. 2, we plot the T-statistic of the slope coefficient for each STFT component of each source point in the two ROIs by STFT-R (Fig. 2(b)) and by MNE-R (Fig. 2(c)), and compared them with the true coefficients in Fig. 2(a) ($SNR = 0.5$, *noise level* $= 0.5$). The T-statistics for the real and imaginary parts are shown separately. In Fig. 2(a), (b) and (c), the vertical axis corresponds to the indices of source points, concatenated for the two ROIs. The horizontal axis corresponds to the indices of STFT components, which is a one-dimensional concatenation of the cells of the frequency $\times$ time matrix in Fig. 2(d); 0–24 are 25 time points in 0 Hz, 25–49 in 6.25 Hz, 50–74 in 12.5 Hz, and so on. STFT-R yielded a sparse pattern, where only the lower frequency (0 to 6.25 Hz) components were active, whereas the pattern by MNE-R spread into higher frequencies (index 100–200, 25–50Hz). We also compared the *averaged absolute T*s for each ROI by STFT-R (Fig. 2(e)) and by MNE-R (Fig. 2(f)), with the true pattern in Fig. 2(d), in which we averaged the absolute values of the real and imaginary parts of the true coefficients across the source points in the ROI. Again, STFT-R yielded a sparse activation pattern similar to the truth, where as MNE-R yielded a more dispersed pattern.

Human Face-Learning Experiment. We applied STFT-R and MNE-R on a subset of data from a face-learning study [19], where participants learned to distinguish two categories of computer-generated faces. In each trial, a participant was shown a face, then reported whether it was Category 0 or 1, and got feedback. In about 500 trials, participants' behavioural accuracy rates increased from chance to at least 70 %. Figure 3(a) shows the smoothed behavioral accuracy of one participant for Category 0, where the smoothing was done by a logistic regression on Legendre polynomial basis functions of degree 5. We used face-selective ROIs pre-defined in an independent dataset, and applied STFT-R and MNE-R to regress on the smoothed accuracy rates. Considering that the participants might have different learning rates for different categories, we analyzed trials with each category separately. Again, it was a simple linear regression with only one covariate, where we fit a slope and an intercept for each STFT component, and we were mainly interested in the slope regression coefficients, which reflected how neural signals correlated with learning. We preprocessed the sensor data using MNE-python and re-sampled the data at 100 Hz. STFT was computed in a time window of 160 ms, at a step size $\tau_0 = 40$ ms. When applying STFT, we set the weights of α for the ROI groups to zero, and used equal weights for other non-ROI groups, which summed to 1. All of the tuning parameters in both methods, including α, β and γ, were selected via cross-validation. We used $B = 20$ in bootstrapping.

We report here preliminary results in one of the face selective regions, the right inferior occipital gyrus (labelled as IOG_R-rh), for one participant and one face category. This area is part of the "occipital face area" reported in the

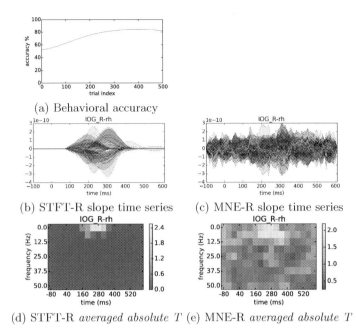

(a) Behavioral accuracy

(b) STFT-R slope time series   (c) MNE-R slope time series

(d) STFT-R *averaged absolute T* (e) MNE-R *averaged absolute T*

Fig. 3. Face-learning experiment results for one participant. (a), Smoothed behavioral accuracy for Category 0. (b) and (c), Time series reconstructed from the STFT slope coefficients in trials with faces of Category 0. Each curve denotes one source point. The shaded bands show 95 % confidence intervals. (d, e) *Averaged absolute T*s by STFT-R and MNE-R.

literature [16]. Since both STFT and the regression on the covariates are linear, we inversely transformed the slope coefficients of the STFT components to a time series for each source point, (denoted by "slope time series"), which showed the slope coefficient in the time domain (Fig. 3(b) and (c)). We observed that STFT-R produced smooth slope time series due to the sparse STFT representation (Fig. 3(b)), whereas MNE-R produced more noisy time series (Fig. 3(c)). We also show the previously defined *averaged absolute T*s in the ROI, produced by STFT-R (Fig. 3(d)) and MNE-R (Fig. 3(e)) Compared with the dispersed pattern by MNE-R, STFT-R produced a more sparse pattern localized near $200 \sim 300$ ms. We speculate that this pattern corresponds to the N250 component near 250 ms, which is related to familiarity of faces [18].

4   Discussion

To estimate learning effects in MEG, we introduced a source localization model (STFT-R), in which we embedded regression of STFT components of source signals on covariates, and exploited a hierarchical L_{21} penalty to induce structured sparsity and emphasize regions of interest. We derived the FISTA algorithm and an active-set strategy to solve STFT-R. In reconstructing the ROI source signals from simulated data, STFT-R achieved smaller errors than a two-step method using MNE, and in a human learning experiment, STFT-R yielded more sparse and thus more interpretable results in identifying what time-frequency components of the ROI signals were correlated with learning. In future work, the STFT-R framework can also be used to regress MEG signals on high-dimensional features of stimuli, where the sparsity-inducing property will be able to select important features relevant to complex cognitive processes.

One limitation of STFT-R is its sparse representation of the non-ROI source points. In our simulation, all of the source points outside the four regions had zero signals, and it was reasonable to represent the two irrelevant regions as sparse source points. However, further simulations are needed to test how well STFT-R behaves when the true signals of the non-ROI source points are more dispersed. It is also interesting to develop a one-step regression model based on Bayesian source localization methods [10, 14], where we can relax the hard sparse constraints but still regularize the problem according to prior knowledge.

Acknowledgements. This work was funded by the Multi-Modal Neuroimaging Training Program (MNTP) fellowship from the NIH (5R90DA023420-08,5R90DA02 3420-09) and Richard King Mellon Foundation. We also thank Yang Xu and the MNE-python user group for their help.

Appendix

Appendix 1

Short-Time Fourier Transform (STFT). Our approach builds on the STFT implemented by Gramfort et al. in [6]. Given a time series $U = \{U(t), t = 1, \cdots, T\}$, a time step τ_0 and a window size T_0, we define the STFT as

$$\Phi(\{U(t)\}, \tau, \omega_h) = \sum_{t=1}^{T} U(t) K(t - \tau) e^{(-i\omega_h)} \tag{6}$$

for $\omega_h = 2\pi h/T_0, h = 0, 1, \cdots, T_0/2$ and $\tau = \tau_0, 2\tau_0, \cdots n_0\tau_0$, where $K(t - \tau)$ is a window function centered at τ, and $n_0 = T/\tau_0$. We concatenate STFT components at different time points and frequencies into a single vector in $V \in \mathbb{C}^s$, where $s = (T_0/2 + 1) \times n_0$. Following notations in [6], we also call the $K(t - \tau) e^{(-i\omega_h)}$ terms STFT dictionary functions, and use a matrix's Hermitian transpose Φ^H to denote them, i.e. $(U^T)_{1 \times T} = (V^T)_{1 \times s}(\Phi^H)_{s \times T}$.

Appendix 2

The Karush-Kuhn-Tucker Conditions. Here we derive the Karush-Kuhn-Tucker (KKT) conditions for the hierarchical L_{21} problem. Since the term $f(z) = \frac{1}{2}\sum_{r=1}^{q}||M^{(r)} - G(\sum_{k=1}^{p} X_k^{(r)} Z_k)\Phi^H||_F^2$ is essentially a sum of squared error of a linear problem, we can re-write it as $f(z) = \frac{1}{2}||b - Az||^2$, where z again is a vector concatenated by entries in Z, b is a vector concatenated by $M^{(1)}, \cdots, M^{(q)}$, and A is a linear operator, such that Az is the concatenated $G(\sum_{k=1}^{p} X_k^{(r)} Z_k) \Phi^H, r = 1, \cdots, q$. Note that although z is a complex vector, we can further reduce the problem into a real-valued problem by rearranging the real and imaginary parts of z and A. Here for simplicity, we only derive the KKT conditions for the real case. Again we use $\{g_1, \cdots, g_h, \cdots, g_N\}$ to denote our ordered hierarchical group set, and λ_h to denote the corresponding penalty for group g_h. We also define diagonal matrices D_h such that

$$D_h(l, l) = \begin{cases} 1 & \text{if } l \in g_h \\ 0 & \text{otherwise} \end{cases} \quad \forall h$$

therefore, the non-zero elements of $D_h z$ is equal to $z|_{g_h}$. With the simplified notation, we re-cast the original problem into a standard formulation:

$$\min_z (\frac{1}{2}||b - Az||_2^2 + \sum_h \lambda_h ||D_h z||_2) \tag{7}$$

To better describe the KKT conditions, we introduce some auxiliary variables, $u = Az, v_h = D_h z$. Then (7) is equivalent to

$$\min_{z,u,v_h} (\frac{1}{2}||b - u||_2^2 + \sum_h \lambda_h ||v_h||_2)$$

$$\text{such that } u = Az, \quad v_h = D_h z, \forall h$$

The corresponding Lagrange function is

$$L(z, u, v_h, \mu, \xi_h) = \frac{1}{2}||b - u||_2^2 + \sum_h \lambda_h ||v_h||_2 + \mu^T (Az - u) + \sum_h \xi_h^T (D_h z - v_h)$$

where μ and ξ_h's are Lagrange multipliers. At the optimum, the following KKT conditions hold

$$\frac{\partial L}{\partial u} = u - b - \mu = 0 \tag{8}$$

$$\frac{\partial L}{\partial z} = A^T \mu + \sum_h D_h \xi_h = 0 \tag{9}$$

$$\frac{\partial L}{\partial v_h} = \lambda_h \partial ||v_h||_2 - \xi_h \ni 0, \forall h \tag{10}$$

where $\partial || \cdot ||_2$ is the subgradient of the L_2 norm. From (8) we have $\mu = u - b$, then (9) becomes $A^T (u - b) + \sum_h D_h \xi_h = 0$. Plugging $u = Az$ in, we can see that the first term $A^T (u - b) = A^T (Az - b)$ is the gradient of $f(z) = \frac{1}{2}||b - Az||_2^2$. For a solution z_0, once we plug in $v_h = D_h z_0$, the KKT conditions become

$$\nabla f(z)_{z=z_0} + \sum_h D_h \xi_h = 0 \tag{11}$$

$$\lambda_h \partial \| \boldsymbol{D}_h \boldsymbol{z}_0 \|_2 - \boldsymbol{\xi}_h \ni 0, \forall h \tag{12}$$

In (12), we have the following according to the definition of subgradients

$$\boldsymbol{\xi}_h = \lambda_h \frac{\boldsymbol{D}_h \boldsymbol{z}_0}{\| \boldsymbol{D}_h \boldsymbol{z}_0 \|_2} \text{ if } \| \boldsymbol{D}_h \boldsymbol{z}_0 \|_2 > 0$$

$$\| \boldsymbol{\xi}_h \|_2 \leq \lambda_h \text{ if } \| \boldsymbol{D}_h \boldsymbol{z}_0 \|_2 = 0$$

Therefore we can determine whether (11) and (12) hold by solving the following problem.

$$\min_{\boldsymbol{\xi}_h} \frac{1}{2} \| \nabla f(\boldsymbol{z})_{\boldsymbol{z}=\boldsymbol{z}_0} + \sum_h \boldsymbol{D}_h \boldsymbol{\xi}_h \|_2^2$$

$$\text{subject to } \boldsymbol{\xi}_h = \lambda_h \frac{\boldsymbol{D}_h \boldsymbol{z}_0}{\| \boldsymbol{D}_h \boldsymbol{z}_0 \|_2} \text{ if } \| \boldsymbol{D}_h \boldsymbol{z}_0 \|_2 > 0$$

$$\| \boldsymbol{\xi}_h \|_2 \leq \lambda_h \text{ if } \| \boldsymbol{D}_h \boldsymbol{z}_0 \|_2 = 0$$

which is a standard group lasso problem with no overlap. We can use coordinate-descent to solve it. We define $\frac{1}{2} \| \nabla f(\boldsymbol{z})_{\boldsymbol{z}=\boldsymbol{z}_0} + \sum_h \boldsymbol{D}_h \boldsymbol{\xi}_h \|_2^2$ at the optimum as a measure of violation of the KKT conditions.

Let f_J be the function f constrained on a set J. Because the gradient of f is linear, if $\boldsymbol{z}_0$ only has non-zero entries in J, then the entries of $\nabla f(\boldsymbol{z})$ in J are equal to $\nabla f_J(\boldsymbol{z}|_J)$ at $\boldsymbol{z} = \boldsymbol{z}_0$. In addition, $\boldsymbol{\xi}_h$'s are separate for each group. Therefore if $\boldsymbol{z}_0$ is an optimal solution to the problem constrained on J, then the KKT conditions are already met for entries in J (i.e. $(\nabla f(\boldsymbol{z})_{\boldsymbol{z}=\boldsymbol{z}_0} + \sum_h \boldsymbol{D}_h \boldsymbol{\xi}_h)|_J = 0$); for $g_h \not\subset J$, we use ($\frac{1}{2} \| (\nabla f(\boldsymbol{z})_{\boldsymbol{z}=\boldsymbol{z}_0} + \sum_h \boldsymbol{D}_h \boldsymbol{\xi}_h)|_{g_h} \|^2$) at the optimum as a measurement of how much the elements in group g_h violate the KKT conditions, which is a criterion when we greedily add groups (see Algorithm 2).

References

1. Bach, F., Jenatton, R., Mairal, J., Obozinski, G.: Optimization withsparsity-inducing penalties. CoRR abs/1108.0775 (2011). http://arXiv.org/abs/1108.0775
2. Beck, A., Teboulle, M.: A fast iterative shrinkage-thresholding algorithm for linear inverse problems. SIAM J. Imaging Sci. **2**(1), 183–202 (2009)
3. Dale, A.M., Liu, A.K., Fischl, B.R., Buckner, R.L., Belliveau, J.W., Lewine, J.D., Halgren, E.: Dynamic statistical parametric mapping: combining fmri and meg for high-resolution imaging of cortical activity. Neuron **26**(1), 55–67 (2000)
4. Galka, A., Ozaki, O.Y.T., Biscay, R., Valdes-Sosa, P.: A solution to the dynamical inverse problem of eeg generation using spatiotemporal kalman filtering. NeuroImage **23**, 435–453 (2004)
5. Gauthier, I., Tarr, M.J., Moylan, J., Skudlarski, P., Gore, J.C., Anderson, A.W.: The fusiform face area is part of a network that processes faces at the individual level. J. Cogn. Neurosci. **12**(3), 495–504 (2000)
6. Gramfort, A., Strohmeier, D., Haueisen, J., Hamalainen, M., Kowalski, M.: Time-frequency mixed-norm estimates: sparse M/EEG imaging with non-stationary source activations. NeuroImage **70**, 410–422 (2013)

7. Gramfort, A., Luessi, M., Larson, E., Engemann, D.A., Strohmeier, D., Brodbeck, C., Parkkonen, L., Hmlinen, M.S.: MNE software for processing MEG and EEG data. NeuroImage **86**, 446–460 (2014)
8. Hamalainen, M., Ilmoniemi, R.: Interpreting magnetic fields of the brain: minimum norm estimates. Med. Biol. Eng. Comput. **32**, 35–42 (1994)
9. Hamalainen, M., Hari, R., Ilmoniemi, R.J., Knuutila, J., Lounasmaa, O.V.: Magnetoencephalography-theory, instrumentation, to noninvasive studies of the working human brain. Rev. Mod. Phys. **65**, 414–487 (1993)
10. Henson, R.N., Wakeman, D.G., Litvak, V., Friston, K.J.: A parametric empirical bayesian framework for the EEG/MEG inverse problem: generative models for multi-subject and multi-modal integration. Front. Hum. Neurosci. **5**, 76 (2011)
11. Jenatton, R., Mairal, J., Obozinski, G., Bach, F.: Proximal methods for hierarchical space coding. J. Mach. Learn. Res. **12**, 2297–2334 (2011)
12. Kanwisher, N., McDermott, J., Chun, M.M.: The fusiform face area: a module in human extrastriate cortex specialized for face perception. J. Neurosci. **17**(11), 4302–4311 (1997)
13. Lamus, C., Hamalainen, M.S., Temereanca, S., Brown, E.N., Purdon, P.L.: A spatiotemporal dynamic distributed solution to the MEG inverse problem. NeuroImage **63**, 894–909 (2012)
14. Mattout, J., Phillips, C., Penny, W.D., Rugg, M.D., Friston, K.J.: MEG source localization under multiple constraints: an extended Bayesian framework. NeuroImage **30**(3), 753–767 (2006)
15. Pascual-Marqui, R.: Standardized low resolution brain electromagnetic tomography (sLORETA): technical details. Methods Find. Exp. Clin. Pharmacol. **24**, 5–12 (2002)
16. Pitcher, D., Walsh, V., Duchaine, B.: The role of the occipital face area in the cortical face perception network. Exp. Brain Res. **209**(4), 481–493 (2011)
17. Stine, R.A.: Bootstrp prediction intervals for regression. J. Am. Stat. Assoc. **80**, 1026–1031 (1985)
18. Tanaka, J.W., Curran, T., Porterfield, A.L., Collins, D.: Activation of preexisting and acquired face representations: the N250 event-related potential as an index of face familiarity. J. Cogn. Neurosci. **18**(9), 1488–1497 (2006)
19. Xu, Y.: Cortical spatiotemporal plasticity in visual category learning (doctoral dissertation) (2013)

Causality and Time-Series

Classification-Based Causality Detection in Time Series

Danilo Benozzo[1,2(✉)], Emanuele Olivetti[1,2], and Paolo Avesani[1,2]

[1] NeuroInformatics Laboratory (NILab), Bruno Kessler Foundation, Trento, Italy
[2] Center for Mind and Brain Sciences (CIMeC), University of Trento, Trento, Italy
{benozzo,olivetti,avesani}@fbk.eu

Abstract. Brain effective connectivity aims to detect causal interactions between distinct brain units and it can be studied through the analysis of magneto/electroencephalography (M/EEG) signals. Methods to evaluate effective connectivity belong to the large body of literature related to detecting causal interactions between multivariate autoregressive (MAR) data, a field of signal processing. Here, we reformulate the problem of causality detection as a supervised learning task and we propose a classification-based approach for it. Our solution takes advantage of the MAR model by generating a labeled data set that contains trials of multivariate signals for each possible configuration of causal interactions. Through the definition of a proper feature space, a classifier is trained to identify the causality structure within each trial. As evidence of the efficacy of the proposed method, we report both the cross-validated results and the details of our submission to the causality detection competition of Biomag2014, where the method reached the 2nd place.

1 Introduction

A main part of neuroscience research concerns brain connectivity and aims to investigate the pattern of interactions between distinct units within the brain [10]. The concept of brain units is strongly related to the level of the adopted scale. Thus, brain connectivity can be studied from the microscopic level of single synaptic connections to the macroscopic level of brain regions. Depending on the type of interactions that we focus on, the topic of brain connectivity is divided into structural, functional and effective connectivity. In the first case the connectivity patterns are referred to anatomical links i.e. neural pathways, in the second case to the statistical dependences between brain activity in different units and in the last one to the causal interactions between them [15].

In particular, effective connectivity provides information about the direct influence that one or more units exert over another and aims to establish causal interactions among them [7]. To achieve this goal the usefulness of brain signals measured by magneto/electroencephalography (M/EEG) has been largely shown [3]. In fact, M/EEG record high temporal resolution signals that directly measure the brain activity. A large body of work was developed about methods to quantify the effective connectivity, mainly in the field of signal processing

© Springer International Publishing AG 2016
I. Rish et al. (Eds.): MLINI 2014, LNAI 9444, pp. 85–93, 2016.
DOI: 10.1007/978-3-319-45174-9_9

where it is known as the problem of *inferring causality among time series*. An overview of the literature is provided below.

A first distinction that can be made in the available methods for causality detection, is between linear and nonlinear methods. Linear approaches are largely used both in time and frequency domain. An example of time domain technique is the Granger causality index. Granger causality is one of the most widespread measure to estimate the direction of causal influence in time series and its basic assumption, that a cause has to precede its effect, has been adopted in many other methods [8]. More precisely, if one or more time series $x_0(t), \ldots, x_k(t)$ are causing the time series $y(t)$, then a future value of $y(t)$ is better predicted by considering also the past values of $x_0(t), \ldots, x_k(t)$ than only those of $y(t)$. Most of the other time domain methods have the property that their multivariate extension is based on the partial auto- and cross-spectra estimation done by frequency-domain methods [16]. Thus, these latter have great adoption in causality assessment [5]. Examples are: the direct transfer function (DTF) [11,12], the direct coherence (DC) [2] and the partial direct coherence [1].

In situations in which the nonlinear component of the causal interaction is expected to be important, nonlinear multivariate methods are used [14]. A first attempt to deal with nonlinearity was done by the local application of linear multivariate methods in order to perform nonlinear prediction [6]. Further approaches are based on information theory [9], phase synchronization [4] and state space synchronization [13].

The intricate structure of interconnections, the enormous amount of dependence that brain units can exert over each other and, last but not least, the lack of a ground truth, make the assessment of the causal interactions a very complex problem. In general new methods to estimate causal interactions are assessed and validated on a limited set of signals and often by using data simulated by multivariate autoregressive (MAR) model. This is a common premise that allows researchers to analyse the performance of their techniques in the fully controlled environment of the MAR model. An example of the interest that has been addressed to causality in multivariate time series is the Biomag2014 Causality Challenge (Causal2014)[1]. The purpose of the contest was to estimate the direct causal interactions in a data set of simulated trials. One trial is meant as three multivariate time series, generated by a known MAR model, that is expected to simulate the behaviour of three neuronal populations.

In this paper we propose a new approach for the causality detection in time series by attacking the problem from a different prospective. Instead of developing a solution in the context of signal processing, as in the previous literature, we faced the problem from the machine learning point of view. Since modelling causal interactions with a MAR model is a common practice in the literature, we used the competition MAR model to create a set of trials for each possible causal configuration among the time series. Then a classifier was trained on those data in order to

---

[1] http://www.biomag2014.org/competition.shtml, see "**Challenge 2**: Causality Challenge".

discriminate between causal configurations. Finally, it was applied to the competition data set providing a solution that reached the second place of Causal2014.

2 Materials

The competition organizers provided the code of the MAR model together with the data set of which to estimate the direct causal interactions. Here, we will describe them both.

The final output of the MAR model is the multivariate time series $\mathbf{X} = \{X(t), t = 0, 1, \ldots, N-1\}, X(t) \in \mathbb{R}^{M \times 1}$ that is defined as the linear combination of two M-dimensional multivariate time series $\mathbf{X_s}$ and $\mathbf{X_n}$

$$\mathbf{X} = (1 - \gamma)\mathbf{X_s} + \gamma \mathbf{X_n} \tag{1}$$

$\mathbf{X_s}$ carries the causal information, $\mathbf{X_n}$ represents the noise corruption and $\gamma \in [0, 1]$ tunes the signal-to-noise ratio. Each time point of $\mathbf{X_s}$ and $\mathbf{X_n}$ is computed by following the MAR model

$$
\begin{aligned}
X_s(t) &= \sum_{\tau=1}^{\min(P,t)} A_s^{(\tau)\top} X_s(t - \tau) + \varepsilon_s(t) \\
X_n(t) &= \sum_{\tau=1}^{\min(P,t)} A_n^{(\tau)\top} X_n(t - \tau) + \varepsilon_n(t)
\end{aligned}
\tag{2}
$$

where P is the order of the MAR model and represents the maximal time lag. $\varepsilon_s(t)$ and $\varepsilon_n(t)$ are realizations from a M-dimensional standard normal distribution. And $A_s^{(\tau)}, A_n^{(\tau)} \in \mathbb{R}^{M \times M}, \tau = 1, 2, \ldots, P$ are the coefficient matrices modelling the influence of the signal values at time $t - \tau$ on the current signal values, i.e. at time t. The coefficient matrices $\{A_s^{(\tau)}\}_\tau$ are involved in the process of causal-informative data generation. They are computed by randomly corrupting the non-zero elements of the $M \times M$ binary matrix A, called configuration matrix. In essence, the configuration matrix A contains the causal structure that leads the MAR model. Specifically $A_{i,j} = 1$ means signal i causes the signal j. On the other hand, coefficient matrices $A_n^{(\tau)}$ lead the noisy part of the signals and they are obtained by randomly generating P diagonal matrices. The diagonality of these latter matrices is needed to avoid noise regressive dependencies across signals. After that, if both sets of matrices $A_s^{(\tau)}$ and $A_n^{(\tau)}$ fulfil the stationarity condition, each time point of $\mathbf{X_s}$ and $\mathbf{X_n}$ can be generated by Eq. 2.

In essence, given P, γ and A, it is possible to generate $\mathbf{X}$ following Eqs. 1 and 2. The goal of the competition is to reconstruct A given $\mathbf{X}$.

The competition data set was built by generating 1000 trials with the following parameter assignments: the number of time series in each trial is $M = 3$, the MAR model order is $P = 10$ and the time series length is $N = 6000$. The trial-specific parameters γ and A were randomly sampled from a standard uniform distribution for each trial and kept secret by the organizer of the competition. From now on, we will refer to the competition data set as $\mathbf{C}$.

3   Methods

The solution that we propose to the causality detection problem is based on a
supervised approach. Indeed, this task to reconstruct A from the data can be
formulated as a classification problem. In a general setting, each trial is composed by M time series and the final goal is to estimate its $M \times M$ binary
configuration matrix A. Thus, there are $M(M-1)$ free binary parameters and
$2^{M(M-1)}$ possible causal configurations[2].

Our supervised approach aims to train a classifier in order to discriminate
between trials that were generated by different configuration matrices. And since
we aim to predict A given a trial, the classifier is going to treat A as the trial's
class label.

The training of the classifier is done on a new simulated data set generated
by the MAR model described in Sect. 2. This new data set, that we will call $\mathbf{L}$,
is meant to better represent the entire population of causal configurations that
can be obtained by the adopted model. Therefore, $\mathbf{L}$ contains multiple trials for
each of the possible $2^{M(M-1)}$ causal configurations.

Before the training, a proper feature space has to be defined in order to
extract the causal structure that led the generation of the trial. And once $\mathbf{L}$ has
been mapped on that feature space, a classifier f is trained on it.

The classifier f and the benefit that the feature space provides, are evaluated by estimating the discriminative power of f through cross-validation. The
discriminative power can be maximized by trying different types of classifiers,
by tuning the related parameters and also by adjusting the feature space. Such
way of proceeding does not introduce circularity because we are not using $\mathbf{C}$.

In the end, f is applied to the competition data set $\mathbf{C}$ to predict the configuration matrix of each trial.

The feature space that we built, is strongly based on the concept of Granger
causality. Indeed, it is a collection of measures that quantifies the ability to
predict the value at a given time point of a certain time series (effect) from the
past values of each possible subset of the M time series in the trial (causes).
The pair, made by causes and effect, is called causality scenario and, for M time
series, there are $\sum_{i=1}^{M} \binom{M}{i} M$ scenarios. In the case of the competition, where
$M = 3$, the possible causality scenarios are 21, and they are summarized in
Table 1, where $x_i(t), i = 0, 1, 2$, denotes each of the time series that defines a
trial.

For each causality scenario, a plain linear regression problem was built by
selecting, as dependent variable, a set of time points from the signal in the *effect*
column. Each of these dependent variables has a regressor vector composed by
the P previous time points selected from the signals in the *causes* column. Table 2
shows how the regression problems were defined when $M = 3$, by specifying
from which time series and time points, regressors and dependent variables are
extracted. In the following, in order to simplify the notation, we will use x_i^t
instead of $x_i(t)$, $i = 0, 1, 2$ and $t \in \mathbf{T}, \mathbf{T} \subseteq \{P, P+1, \ldots, N-1\}$. Figure 1 explains

---

[2] The diagonal is not relevant since by definition the time series are autoregressive.

Table 1. The possible causality scenarios for three time series $x_i(t), i = 0, 1, 2$.

Causes	Effect
$x_0(t)$	$x_i(t)$
$x_1(t)$	$x_i(t)$
$x_2(t)$	$x_i(t)$
$x_0(t), x_1(t)$	$x_i(t)$
$x_0(t), x_2(t)$	$x_i(t)$
$x_1(t), x_2(t)$	$x_i(t)$
$x_0(t), x_1(t), x_2(t)$	$x_i(t)$

how, for the specific time point $t = 30$, the input of the regression problem is built in the case of the last causality scenario of the Table 2 with $i = 2$. More precisely, this example shows how the input of the regression problem is defined in order to quantify the plausibility of the causality scenario: "x_0, x_1 and x_2 are causing x_2".

Table 2. Description of how the 21 linear regression problems are defined for each trial. x_i^t, $i = 0, 1, 2$ and $t \in \mathbf{T}, \mathbf{T} \subseteq \{10, 11, \dots, N - 1\}$, are the three time series of a trial.

Regressors (causes)	Dependent variable (effect)
$[x_0^{t-10}, \dots, x_0^{t-1}]$	x_i^t
$[x_1^{t-10}, \dots, x_1^{t-1}]$	x_i^t
$[x_2^{t-10}, \dots, x_2^{t-1}]$	x_i^t
$[x_0^{t-10}, \dots, x_0^{t-1}, x_1^{t-10}, \dots, x_1^{t-1}]$	x_i^t
$[x_0^{t-10}, \dots, x_0^{t-1}, x_2^{t-10}, \dots, x_2^{t-1}]$	x_i^t
$[x_1^{t-10}, \dots, x_1^{t-1}, x_2^{t-10}, \dots, x_2^{t-1}]$	x_i^t
$[x_0^{t-10}, \dots, x_0^{t-1}, x_1^{t-10}, \dots, x_1^{t-1}, x_2^{t-10}, \dots, x_2^{t-1}]$	x_i^t

The regression problem of each causality scenario was cross-validated and its performance was quantified through multiple regression metrics, e.g. mean square error. The ensemble of the regression metrics of each causality scenario defined the initial feature vector of the trial. We then applied standard feature engineering techniques on the initial feature vector to enrich the feature space. The choice of using multiple regression metrics and in particular which ones including in the initial feature vectors, as well the choice of the feature engineering techniques, are driven by the goal to maximize the discriminative power of f. See Sect. 4 for further details.

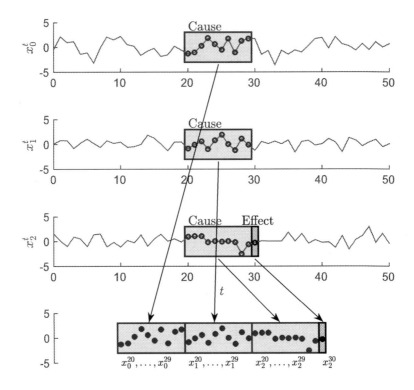

Fig. 1. Example of how the sample associated at the time point $t = 30$ is built in order to form the input of the last regression problem of the Table 2, for the case $i = 2$.

4   Experiments

In this section we present the technical details and results of the experiments that were conducted to evaluate the method described in Sect. 3. In particular, we show two different types of results. The first one is an estimate of the discriminative power of a classifier trained on the **L** data set and it provides a quantification of how well the defined feature space is able to express the causal structure behind a trial. The second result is the competition score obtained by our submission, which gives us insights into how our approach works compared to the ones adopted by the other participants.

Results are presented both in terms of confusion matrices and competition score. The competition score was defined in the following way. For each entry $\hat{A}_{ij}$, $i \neq j$, of each predicted $\hat{A}$, if $\hat{A}_{ij}$ was 1 and correct, then $+1$ point was given. If $\hat{A}_{ij}$ was 1 but incorrect, then -3 points were given. If $\hat{A}_{ij}$ was 0, then 0 points were given. In practice, false discoveries were punished three times more than what true discoveries were rewarded.

In order to take into account the strong false positive penalisation, we added a cost model to our predictions, by combining the probability of each of the 64 classes with the cost of predicting one class instead of another. Given S_{ij} the cost of predicting i when the true class was j, the optimal way to assign the class l to a trial is

$$l = \underset{i=1,2,...,64}{\operatorname{argmax}} \sum_{j=1}^{64} S_{i,j} p_j \qquad (3)$$

where p_j is the probability of class j for the trial, as estimated by the classifier.

The new simulated labeled data set $\mathbf{L}$ was generated by keeping the same parameter initialization[3] of $\mathbf{C}$, except for the number of trials that was increased to 64000 in order to have 1000 trials for each class. Indeed, since $M = 3$ the amount of causal configurations is $2^6 = 64$. The regression metrics used to build the feature space are the mean square error and the coefficient of determination r^2. Both were included because we noticed a significant improvement in the cross-validated score, although, intuitively, they could seem redundant. We also added an estimate of the Granger causality coefficients[4] to the feature space.

As a final step we increased the number of features through standard feature engineering techniques by applying simple basis functions. This consisted in extracting the 2nd power, 3rd power and square root of the previously defined features, together with the pairwise product of all features. Adding extracted features was motivated both by the need to overcome the limitation of the adopted linear classifier and because they proved to be effective in increasing the cross-validated score.

Both the data sets, $\mathbf{L}$ and $\mathbf{C}$, were mapped to the proposed feature space. Then the performance of the logistic regression classifier[5], with ℓ_2 regularisation, was evaluated on $\mathbf{L}$ through 5-folds cross-validation. In this way we quantified the discriminative capability of the proposed method.

Tables 3 and 4 show the cross-validated classification results in $\mathbf{L}$ by means of confusion matrices. In particular, Table 3 is related to the percentage of causal interactions predicted by assigning to each test trial the most probable class, i.e. $l = \operatorname{argmax} p_i$, and its accuracy is 81%. In Table 4 the assignments are done by Eq. 3 according to the cost matrix, i.e. by penalizing the false positives,

Table 3. Confusion matrix computed by assigning to each test trial the most probable class.

		Predicted	
		1	0
True	1	79%	21%
	0	17%	83%

Table 4. Confusion matrix in which the test trial class labels are computed by Eq. 3.

		Predicted	
		1	0
True	1	56%	44%
	0	1%	99%

---

[3] Excluding the trial-specific parameter γ which was randomly uniformly generated for each trail.

[4] http://nipy.org/nitime.

[5] http://scikit-learn.org.

and the related accuracy is 77.5 %. Through their comparison, the effect of S is evident since in Table 4, false positives are strongly decreased, due to the score penalization, but to the detriment of some true positives.

Finally logistic regression was trained on $\mathbf{L}$ and tested on $\mathbf{C}$ to predict the configuration matrices of the competition. According to the number of trials in $\mathbf{C}$ and the assumptions of the generative process, the expected range of the score is $[-9000, 3000]$. The score of our submission was 1571, which reached the 2nd place in the final ranking of the Causal2014 competition.

5   Discussion, Conclusion and Future Works

In this paper, we proposed a new approach to detect causal interactions in multivariate time series. Specifically, we developed a classification-based causality detection method by defining a feature space based on the concept of Granger causality and by exploiting the MAR model as data generator. Aside from the novelty of the method itself, the interesting aspect of our solution is that it is a supervised method. Thus, it belongs to the machine learning field and not to the signal processing as traditionally was for that type of problem.

The proposed method was assessed by cross-validating the generated labeled data set and it provided promising results, as shown in Tables 3 and 4 by means of confusion matrices. Then, the submitted solution to the Causal2014 competition was computed by a classifier trained on the generated labeled data set. The achieved results, both in terms of cross-validation and competition ranking, are evidence that classification-based techniques are a feasible alternative to the signal processing methods for inferring causality between time series. And furthermore, that the defined feature space is able to well capture the causal structures among signals.

As an improvement of our approach, we are working on a tractable extension to the case of detecting causality in more than three time series.

References

1. Baccalà, L.A., Sameshima, K.: Partial directed coherence: a new concept in neural structure determination. Biol. Cybern. **84**(6), 463–474 (2001). http://view.ncbi.nlm.nih.gov/pubmed/11417058
2. Baccalà, L.A., Sameshima, K., Ballester, G., Do Valle, A.C., Timo-Iaria, C.: Studying the interaction between brain structures via directed coherence and granger causality. Appl. Signal Process. **5**, 40–48 (1998). http://www.lcs.poli.usp.br/~baccala/pdc/papers/asp.pdf
3. Brookes, M.J., Woolrich, M.W., Barnes, G.R.: Measuring functional connectivity in MEG: a multivariate approach insensitive to linear source leakage. NeuroImage **63**(2), 910–920 (2012). http://view.ncbi.nlm.nih.gov/pubmed/22484306
4. Butler, S.R., Glass, A.: Asymmetries in the electroencephalogram associated with cerebral dominance. Electroencephalogr. Clin. Neurophysiol. **36**(5), 481–491 (1974). http://view.ncbi.nlm.nih.gov/pubmed/4135345

5. Faes, L., Erla, S., Nollo, G.: Measuring connectivity in linear multivariate processes: definitions, interpretation, and practical analysis. Comput. Math. Methods Med. **2012**, 1–18 (2012). doi:10.1155/2012/140513
6. Freiwald, W.A., Valdes, P., Bosch, J., Biscay, R., Jimenez, J.C., Rodriguez, L.M., Rodriguez, V., Kreiter, A.K., Singer, W.: Testing non-linearity and directedness of interactions between neural groups in the macaque inferotemporal cortex. J. Neurosci. Methods **94**(1), 105–119 (1999). http://view.ncbi.nlm.nih.gov/pubmed/10638819
7. Friston, K.J.: Functional and effective connectivity: a review. Brain Connectivity **1**(1), 13–36 (2011). doi:10.1089/brain.2011.0008
8. Granger, C.W.J.: Investigating causal relations by econometric models and cross-spectral methods. Econometrica **37**(3), 424–438 (1969). doi:10.2307/1912791
9. Hlavackovaschindler, K., Palus, M., Vejmelka, M., Bhattacharya, J.: Causality detection based on information-theoretic approaches in time series analysis. Phys. Rep. **441**(1), 1–46 (2007). doi:10.1016/j.physrep.2006.12.004
10. Horwitz, B.: The elusive concept of brain connectivity. NeuroImage **19**(2 Pt 1), 466–470 (2003). http://view.ncbi.nlm.nih.gov/pubmed/12814595
11. Kamiński, M., Ding, M., Truccolo, W.A., Bressler, S.L.: Evaluating causal relations in neural systems: granger causality, directed transfer function and statistical assessment of significance. Biol. Cybern. **85**(2), 145–157 (2001). http://view.ncbi.nlm.nih.gov/pubmed/11508777
12. Kaminski, M.J., Blinowska, K.J.: A new method of the description of the information flow in the brain structures. Biol. Cybern. **65**(3), 203–210 (1991). doi:10.1007/bf00198091
13. Papana, A., Kugiumtzis, D., Larsson, P.G.: Reducing the bias of causality measures. Phys. Rev. E **83**(3) (2011). http://dx.doi.org/10.1103/physreve.83.036207
14. Pereda, E., Quiroga, R.Q.Q., Bhattacharya, J.: Nonlinear multivariate analysis of neurophysiological signals. Prog. Neurobiol. **77**(1–2), 1–37 (2005). doi:10.1016/j.pneurobio.2005.10.003
15. Sakkalis, V.: Review of advanced techniques for the estimation of brain connectivity measured with EEG/MEG. Comput. Biol. Med. **41**(12), 1110–1117 (2011). doi:10.1016/j.compbiomed.2011.06.020
16. Winterhalder, M., Schelter, B., Hesse, W., Schwab, K., Leistritz, L., Klan, D., Bauer, R., Timmer, J., Witte, H.: Comparison of linear signal processing techniques to infer directed interactions in multivariate neural systems. Sig. Process. **85**(11), 2137–2160 (2005). doi:10.1016/j.sigpro.2005.07.011

Fast and Improved SLEX Analysis of High-Dimensional Time Series

Ahmed Hefny[1(✉)], Robert E. Kass[1], Sanjeev Khanna[2], Matthew Smith[2], and Geoffrey J. Gordon[1]

[1] Carnegie Mellon University, Pittsburgh, USA
ahefny@cs.cmu.edu
[2] University of Pittsburgh, Pittsburgh, USA

Abstract. We address the problem of segmenting a multi-dimensional time series into stationary blocks by improving AutoSLEX [1], which has been successfully used for this purpose. AutoSLEX finds the best basis in a library of smoothed localized exponentials (SLEX) basis functions that are orthogonal and localized in both time and frequency. We introduce DynamicSLEX, a variant of AutoSLEX that relaxes the dyadic intervals constraint of AutoSLEX, allowing for more flexible segmentation while maintaining tractability. Then, we introduce RandSLEX, which uses random projections to scale-up SLEX-based segmentation to high dimensional inputs and to establish a notion of strength of splitting points in the segmentation. We demonstrate the utility of the proposed improvements on synthetic and real data.

Keywords: SLEX · Time series

1 Introduction

When analyzing neural data such as Electroencephalography (EEG), magnetoencephalography (MEG), local field potential (LFP) or spike trains, it is typical to encounter non-stationary multidimensional time series whose spectra change over time depending on external stimuli and behavioral states of the organism [2]. In such cases, one can assume the signal to be locally stationary and hence, it is desirable to decompose the time series into orthogonal basis functions that are localized in both time and frequency. To achieve these qualities, Ombao et al. [1] proposed the use of smooth localized exponentials (SLEX). The SLEX transform constructs orthogonal basis functions that are localized in time and frequency by applying *two* special window functions to the Fourier bases. The AutoSLEX model creates a library of SLEX basis functions that correspond to dyadic time intervals of decreasing length. A segmentation of a time series into stationary segments can be obtained by selecting the best basis from the library using the best basis algorithm (BBA) [3]. The choice of dyadic intervals makes the segmentation problem tractable. However, it limits the obtainable segmentations. More recent methods relaxed that limitation, at the expense of

© Springer International Publishing AG 2016
I. Rish et al. (Eds.): MLINI 2014, LNAI 9444, pp. 94–103, 2016.
DOI: 10.1007/978-3-319-45174-9_10

having an intractable problem for which MCMC based methods are employed
[4]. After reviewing SLEX in Sect. 2. We introduce two improvements to SLEX-
based segmentation: in Sect. 3 we propose DynamicSLEX, a variant of SLEX
that overcomes the dyadic intervals limitation of AutoSLEX while maintaining
tractability; and in Sect. 4 we propose RandSLEX, a method to scale-up SLEX
analysis to high-dimensional inputs using random projections.

2 Background

2.1 SLEX Transform

SLEX basis functions are obtained by applying special pairs of window functions
to the Fourier basis. A SLEX basis function has the form

$$\phi_\omega(t) = \Psi_+(t)\exp(i2\pi\omega t) + \Psi_-(t)\exp(-i2\pi\omega t), \tag{1}$$

where $\omega \in [-1/2, 1/2]$. The window functions Ψ_+ and Ψ_- are parametrized by an
interval $[\alpha_0, \alpha_1]$ and overlap ϵ and have compact support on $[\alpha_0 - \epsilon, \alpha_1 + \epsilon]$.[1] In
more detail, Ψ_+ and Ψ_- are given by

$$\Psi_+(t) = r^2\left(\frac{t-\alpha_0}{\epsilon}\right)r^2\left(\frac{t-\alpha_1}{\epsilon}\right)$$

$$\Psi_-(t) = r\left(\frac{t-\alpha_0}{\epsilon}\right)r\left(\frac{\alpha_0-t}{\epsilon}\right) - r\left(\frac{t-\alpha_1}{\epsilon}\right)r\left(\frac{\alpha_1-t}{\epsilon}\right),$$

where $r(.)$ is the iterated sine function given by

$$r_0(t) = \sin\left[\frac{\pi}{4}(1+t)\right]$$

$$r_{d+1}(t) = r_d\left(\sin\frac{\pi}{2}t\right).$$

The choice of d controls the steepness in the rising and falling phases of Ψ and
hence time-frequency localization properties. Note that if $\Psi_- \equiv 0$ we revert to
STFT. The existence of the proper Ψ_- is what makes the SLEX basis orthogonal
and localized. Figure 1 depicts sample realizations of Ψ_+ and Ψ_-.

A SLEX library is a collection of bases, each of which consists of localized
SLEX waveforms that span dyadic time intervals. Figure 2(b), demonstrates this
idea; each block consists of all waveforms that span the corresponding interval.
Similar to [1], we use $S(j, b)$ to refer to the bth block in the jth level where $S(0, 0)$
is the root block that spans the entire time series. Each choice of non-overlapping
blocks that span the time series corresponds to a basis and a segmentation (see
Fig. 2(c)). The SLEX coefficients of block $S(j, b)$ for an input series X can be
computed as follows:

$$d_{j,b}(\omega_k) = \frac{1}{\sqrt{M_j}}\sum_t X(t)\overline{\phi_{j,b,\omega_k}(t)}, \tag{2}$$

---

[1] In the rest of the paper when we speak of a basis function over some time interval,
we mean $[\alpha_0, \alpha_1]$ (ignoring the overlap).

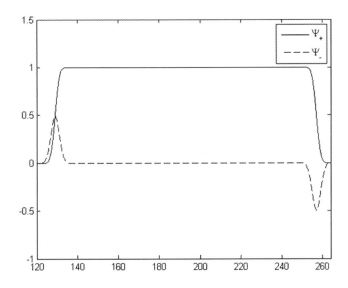

Fig. 1. Ψ_+ and Ψ_- constructed with $\alpha_0 = 128$, $\alpha_1 = 256$ and $\epsilon = 8$

where $M_j = |S(j,b)| = \frac{T}{2^j}$ and ϕ_{j,b,ω_k} is defined as in (1) for window functions Ψ_+ and Ψ_- that correspond to the block $S(j,b)$. Note that the sum over t corresponds to applying a window function followed by inner product with a Fourier basis function. Therefore $d_{j,b}(\omega_k)$ can be computed for all $\omega_k = k/M_j$ for $k = -M_j/2+1,\ldots,M_j/2$ in $O(T \log T)$ time using fast Fourier transform (FFT).

2.2 Basis Selection

To select a SLEX basis, we define a cost function $C(j,b)$ for each block

$$C(j,b) = \sum_{k=-M_j/2+1}^{M_j/2} \log I_{j,b}(\omega_k) + \beta_j \sqrt{M_j} \tag{3}$$

where β_j is a penalty parameter that prevents oversegmentation and $I_{j,b}(\omega_k)$ is obtained by smoothing the SLEX periodogram $|d_{j,b}(\omega_k)|^2$ using a kernel smoother. The smoothing bandwidth is chosen by generalized cross validation (GCV), as detailed in [1]. After computing the costs, basis selection is performed using best basis algorithm [3] which simply proceeds bottom up, merging two sibling blocks if the cost of their parent is less than the sum of their costs.

2.3 The Multivariate SLEX

The straightforward extension of SLEX transform to multiple time series is to assume they are independent and consequently assume that the net cost of a block is the sum of costs of all series [1,3]. To ensure that spectral information is non-redundant in the presence of correlation, Ombao et al. [5] propose computing the cost based on the eigen spectrum of the smoothed cross-periodogram matrix given by

$$I_{j,b}^{(i,j)}(\omega_k) = d_{j,b}^{(X_i)}(\omega_k)\overline{d_{j,b}^{(X_j)}(\omega_k)}, \tag{4}$$

The proposed multivariate cost function is given by

$$C(j,b) = \sum_{k=-M_j/2+1}^{M_j/2} \sum_{d=1}^{P} \frac{\lambda_{j,b}^d(\omega_k) \log \lambda_{j,b}^d(\omega_k)}{\sum_{d'=1}^{P} \lambda_{j,b}^{d'}(\omega_k)} + \beta_j \sqrt{M_j}, \tag{5}$$

where P is the number of time series and $\lambda^d(\omega_k)$ is the dth eigenvalue of the smoothed cross-periodogram matrix at frequency ω_k. This can be thought of as applying the additive cost method on non-stationary principal components that have zero coherence. With the new formulation, basis selection can proceed as described in 2.2. The multivariate SLEX transform can be summarized as follows:

1. For each block $S(j,b)$, compute the SLEX coefficients based on (2) using FFT. Use the coefficients to compute the cross-periodogram matrix for each block based on (4).
2. For each block $S(j,b)$, smooth the cross-periodograms $I_{j,b}^{(i,j)}(\omega_k)$ along frequency using a window smoother whose bandwidth is optimized based on GCV.
3. For each block $S(j,b)$, compute the eigenvalues of the smoothed cross-periodogram matrix at each frequency and use them to compute the cost based on (5).
4. Use BBA to obtain best segmentation and extract the smoothed cross-periodogram matrices corresponding to selected blocks.

3   Flexible Segmentation Using DynamicSLEX

Although AutoSLEX analysis provides an appealing segmentation method, it suffers from a fundamental limitation; if the best basis contains the interval $[2m\frac{T}{2^k}, 2(m+1)\frac{T}{2^k}]$ then it must contain its sibling in the dyadic structure as well as the sibling of the parent, the sibling of the grand parent and so on up to $[0, \frac{T}{2}]$ and $[\frac{T}{2}, T]$. Therefore, AutoSLEX can result in spurious splits of the time series, as demonstrated in Fig. 2, or otherwise miss splitting points. To mitigate this problem, we introduce DynamicSLEX, a variant of SLEX analysis that is also tractable but, at the same time, is capable of producing *any* segmentation whose splitting points are located at integer multiples of $\frac{T}{2^K}$ regardless of the length and starting position of each segment. DynamicSLEX uses the same dyadic bottom-up strategy to select basis functions using BBA. Recall that, in AutoSLEX, a single BBA step determines the best segmentation of a block $S(j,b)$ given the best segmentations of it children $S(j+1,2b)$ and $S(j+1,2b)$

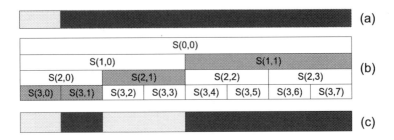

Fig. 2. Limitation of SLEX-based segmentation: to recover the left (bright) segment in (a), bba needs to choose the shaded basis intervals in (b) which produce the segmentation shown in (c).

(denoted in Fig. 3 as "input"). BBA for AutoSLEX can decide to *merge* the blocks, removing all splitting points or to *split* the blocks, keeping their splitting points and introducing a middle one that separates them. DynamicSLEX introduces a third option, to *concatenate* the blocks by keeping the splitting points intact. This is demonstrated in Fig. 3. Effectively, DynamicSLEX is able to merge between adjacent blocks that have different parents and possibly are at different levels, preventing unnecessary splits. With proper bookkeeping, it is easy to compute the cost of blocks resulting concatenation without affecting the tractability of BBA. The algorithm is outlined in Algorithm 1. The algorithm makes use of $ComputeIntervalCost$ function, which computes the cost of a given interval according to (3) or its multivariate version (5)[2].

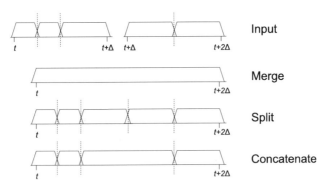

Fig. 3. The effect of merging, split and concatenation on two adjacent intervals. The vertical dashed lines indicate splitting points whereas the window functions indicate the spans of the selected basis functions.

---

[2] Note that concatenation can introduce intervals whose length is not be a power of 2. Therefore, we cannot use the common Cooley-Tukey method for FFT. Instead, to efficiently compute SLEX periodograms in $O(T \log T)$ time, we resort to Chirp-Z Transform [6].

input :
    X: A $T \times P$ matrix representing P time series of length T
    J: Maximum segmentation level
output:
    A set of splitting points
data :
    C(j,b): The cost of block b at level j
    S(j,b): Set of splitting points in block b at level j
    R(j,b): The cost of the rightmost segment in block b at level j
    L(j,b): The cost of the leftmost segment in block b at level j
{*Initialization*}
for $b \leftarrow 1$ **to** 2^J **do**
    $C(J,b) \leftarrow ComputeIntervalCost([(l-1)\frac{T}{2^J}, l\frac{T}{2^J}])$;
    $S(J,b) \leftarrow \{\}$;
    $R(J,b) \leftarrow C(j,b)$; $L(J,b) \leftarrow C(j,b)$;
end
{*Dynamic BBA*}
for $j \leftarrow J$ **to** 0 **do**
    for $b \leftarrow b$ **to** $2^j - 1$ **do**
        $merge\_cost \leftarrow ComputeIntervalCost([b\frac{T}{2^j}, (b+1)\frac{T}{2^j}])$;
        $split\_cost \leftarrow C(j+1, 2b) + C(j+1, 2b+1)$;
        {*Construct interval for new basis function*}
        $left \leftarrow \max(S(j+1, 2b) \cup \{b\frac{T}{2^k}\})$;
        $right \leftarrow \min(S(j+1, 2b+1) \cup \{(b+1)\frac{T}{2^k}\})$;
        $cat\_cost \leftarrow split\_cost - R(j+1, 2b) - L(j+1, 2b+1) +$
        $ComputeIntervalCost([left, right])$;

        {*Select minimum cost*}
        $C(j,b) \leftarrow \min(\{merge\_cost, split\_cost, cat\_cost\})$;
        if $C(j,b) = merge\_cost$ **then** $S(j,b) \leftarrow \{\}$ **else if** $C(j,b) = split\_cost$
        then $S(j,b) \leftarrow S(j+1, 2b) \cup S(j+1, 2b+1) \cup \{(b+\frac{1}{2})\frac{T}{2^j}\}$ **else**
        $S(j,b) \leftarrow S(j+1, 2b) \cup S(j+1, 2b+1)$ Update $R(j,b)$ and $L(j,b)$;
    end
end
Output $S(0,0)$;

Algorithm 1: DynamicSLEX Segmentation Algorithm

4   Fast SLEX Analysis Using RandSLEX

One problem with AutoSLEX is that it does not scale well to high-dimensional time series. For a P–dimensional series, the cost of computation and smoothing of the cross periodograms is $O(P^2T)$ and the cost of eigenvalue decomposition of the cross periodograms is $O(P^3T)$. This cost can be substantial or even prohibitive for the analysis of massive datasets of high dimensional time series. We propose the use of random projections to speed up SLEX analysis. Specifically, we choose $p \ll P$ and generate a p-dimensional time series by taking random Gaussian-distributed linear combinations of the input series X. The use of random projection for dimensionality reduction was successful in numerous applications [7,8]. In our case, the use of random projection is motivated by the observation that a random Gaussian combination of a piecewise-stationary multivariate timeseries preserves stationarity break points almost surely.

Since $p \ll P$, SLEX segmentation will be much faster and can be repeated k times with different randomly generated combinations, potentially resulting in different splitting points. The results of multiple segmentations can be aggregated in a split-count graph as shown in Fig. 4. The split-count graph is a visual summary that gives, for each position in the time series, the number of times that RandSLEX detected a splitting point at that position out of the k different runs. This number can be interpreted as a measure of *strength* for each splitting point and can be used to obtain segmentations at increasing resolutions by filtering out splitting points whose strength is below a given threshold. Once a segmentation is chosen, the coefficients of the full input series X w.r.t the chosen basis can be computed in $O(PT \log T)$ time.

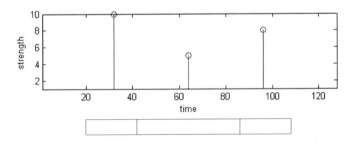

Fig. 4. Top: A sample split-count graph for $k = 10$ runs of RandSLEX. Bottom: segmentation obtained by setting strength threshold to 7.

5   Experiments

5.1   Synthetic Experiments

In the first experiment, we demonstrate the utility of DynamicSLEX. We generated time series data from three auto regressive (AR) processes

$P_1 := AR(1)[0.9]$, $P_2 = AR(2)[1.69, -0.81]$ and $P_3 = AR(3)[1.35, -0.65]$. We
generated a piecewise-stationary series of length $T = 1024$ by switching the
source process at $T/4$, $T/2$, $3T/4$. We experimented with four categories of
series shown in Table 1. Categories (b) and (d) are essentially "harder" versions
of categories (a) and (c), with P_3 less distinguishable from P_2 than P_1. For each
category, we generate 100 time series which are then segmented using AutoSLEX
and DynamicSLEX with minimum block length of $T/4$ and penalty parameter
$\beta = 1$. Table 1 shows the number of times a splitting point was generated at
$T/4$, $T/2$ and $3T/4$. While AutoSLEX and DynamicSLEX are comparable in
categories (a) and (b), AutoSLEX consistently produces a spurious split at $T/2$
in category (c). In category (d) AutoSLEX fails to detect the splits at $T/4$ and
$3T/4$ more often than DynamicSLEX.

Table 1. Categories of time series used to test DynamicSLEX and the number of times
each splitting point is generated by AutoSLEX and DynamicSLEX for each category
across 100 trials.

Category	$[0, T/4]$	$[T/4, T/2]$	$[T/2, 3T/4]$	$[3T/4, T]$	AutoSLEX $T/4$	$T/2$	$3T/4$	DynamicSLEX $T/4$	$T/2$	$3T/4$
(a)	P_1	P_2	P_1		98	99	6	98	99	6
(b)	P_3	P_2	P_3		66	78	5	69	76	6
(c)	P_1	P_2		P_1	99	98	96	98	9	97
(d)	P_3	P_2		P_3	47	52	39	77	12	54

In the second experiment, we asses the ability of RandSLEX to efficiently
recover segmentations of a multivariate time series. For this purpose we generate
four different time series, which switch from process P_1 to process P_2 at times
$T/16$, $T/8$, $5T/16$ and $11T/16$ respectively ($T = 8192$). From these four series,
a 10 dimensional series is obtained by taking random linear combinations of
the base signals where the combinations weights are sampled from a standard
normal distribution. We run RandSLEX on top of DynamicSLEX with $p = 1$ and
$k = 10$ and for each confidence threshold $1 \leq \theta \leq k$ we count true positive, false
positive and false negative splitting points, where a positive splitting point is
one that is reported at least θ times. We aggregate the counts over 100 different
instantiations of that experiment and based on that compute precision (the
percentage of reported splitting points that are true) and recall (the percentage
of true splitting points that were reported) for each confidence threshold θ. The
results are summarized in the PR curve depicted in Fig. 5(left). The curve shows
that, for example, RandSLEX can achieve 94.5 % recall with 81.3 % precision.
Surprisingly, RandSLEX outperformed DynamicSLEX (marked as a green circle
in the figure), which gives 81 % at 41 % recall.

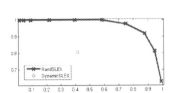 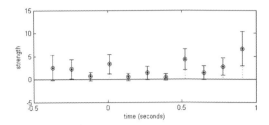

Fig. 5. left: PR curve for RandSLEX on synthetic time series **right:** Mean and standard errors of split strength on LFP data (Color figure online)

5.2 Analysis of Local Field Potential Data

In this experiment, we demonstrate RandSLEX on Local field potentials (LFP) recorded in the Frontal Eye Fields (FEF). Data were acquired through a 16-channel linear probe at 1 kHz from an alert rhesus macaque monkey during performance of a memory-guided saccade task. The animal was required to fixate a central dot during presentation of a visual cue. The visual cue was a brief flash (at time 0.0) at a fixed eccentricity in one of eight directions relative to fixation (spaced by 45°). From time 0.0, the animal had to maintain fixation for a delay period of 500 ms after which the central fixation dot was removed. The animal was rewarded for a successful saccadic eye movement to the remembered direction of the visual cue. The time series, consisting of 1500 samples, was zero-padded to the next power of two. We applied RandSLEX on 100 trials of each direction, for a total of 800 trials. The minimum block size was set to 128 and the parameters p and k set to 1 and 16 respectively. The hypothesis is that changes in the stimulus are expected to change the spectral properties of the time series and cause splitting points. Figure 5(right) summarizes the strength of splitting points across trials (zero padding interval not included). Note that two significant splits were found at 0.0 and 0.5 s, where there is a change in the stimulus. Switching on the fixation dot at around −0.3 s is captured by another less significant splitting point. It is remakable that the model employs no prior knowledge of the stimulus. We speculate that the rightmost splitting point could be attributed to zero padding or other edge effects.

6   Discussion

We introduced an efficient method for segmenting a multivariate into stationary segments. The method provides two enhancements over AutoSLEX: An expanded library of bases that allows for more flexible segmentation (Dynamic-SLEX) and the use of random projections to scale up to high-dimensional time series and provide a notion of segmentation strength. Our experiments revealed that the introduced method gives promising results on synthetic and real data.

References

1. Ombao, H., Raz, J., von Sachs, R., Malow, R.: Automatic statistical analysis of bivariate nonstationary time series. J. Am. Stat. Assoc. **454**, 543–560 (2001)
2. Cranstoun, S.D., Ombao, H.C., von Sachs, R., Guo, W., Litt, B.: Time-frequency spectral estimation of multichannel EEG using the auto-slex method. IEEE Trans. Biomed. Eng. **49**(9), 988–996 (2002)
3. Wickerhauser, M.V.: Adaptive Wavelet-Analysis, Theorie und Software. Vieweg Verlag, Braunschweig/Wiesbaden (1995). German translation of "Adapted Wavelet Analysis from Theory to Software"
4. Rosen, O., Wood, S., Stoffer, D.S.: Adaptspec: Adaptive spectral estimation for nonstationary time series (2012)
5. Ombao, H., von Sachs, R., Guo, W.: SLEX analysis of multivariate nonstationary time series. J. Am. Stat. Assoc. **100**(470), 519–531 (2005)
6. Rabiner, L.R., Schafer, R.W., Rader, C.M.: The chirp z-transform algorithm and its application. Bell Syst. Tech. J. **48**(5), 1249–1292 (1969)
7. Halko, N., Martinsson, P.G., Tropp, J.A.: Finding structure with randomness: probabilistic algorithms for constructing approximate matrix decompositions. SIAM Rev. **53**(2), 217–288 (2011)
8. Bingham, E., Mannila, H.: Random projection in dimensionality reduction: applications to image and text data. In: Proceedings of the Seventh ACM SIGKDD International Conference on Knowledge Discovery and Data Mining, KDD 2001, pp. 245–250. ACM, New York (2001)

Best Paper Awards: MLINI 2013

Predicting Short-Term Cognitive Change from Longitudinal Neuroimaging Analysis

Michael H. Coen[1]([✉]), M. Hidayath Ansari[2], and Barbara B. Bendlin[3]

[1] Department of Biostatistics and Medical Informatics,
University of Wisconsin-Madison, Madison, WI 53706, USA
mhcoen@biostat.wisc.edu
[2] Department of Computer Sciences, University of Wisconsin-Madison,
Madison, WI 53706, USA
[3] Department of Medicine, University of Wisconsin-Madison,
Madison, WI 53792, USA

Abstract. This paper introduces a framework for analyzing longitudinal neuroimaging datasets. We address the problem of detecting subtle, short-term changes in neural structure that are indicative of cognitive decline and correlate with risk factors for Alzheimer's disease. Previous approaches have focused on separating populations with different risk factors based on gross changes, such as decreasing gray matter volume. In contrast, we introduce a new spatially-sensitive kernel that allows us to characterize individuals, as opposed to populations. We use this for both classification and regression, e.g., to predict changes in a subject's cognitive test scores from neuroimaging data alone. In doing so, this paper presents the first evidence demonstrating that very small changes in white matter structure over a two year period can predict change in cognitive function in healthy adults.

1 Introduction

This paper introduces a framework for analyzing *longitudinal* neuroimaging datasets. We address the problem of detecting subtle changes in neural structure that are indicative of cognitive decline and correlate with risk factors for Alzheimer's disease (AD). Previous approaches have focused on separating populations with different risk factors based on gross changes such as decreasing gray and white matter volume [6,8,13] or statistical voxel-based comparisons [3,7]. In contrast, we introduce a new *spatially-sensitive* kernel that allows us to characterize individuals – as opposed to populations – by identifying neural regions that are implicated in cognitive performance. We use this for both classification and regression, e.g., to predict changes in a subject's cognitive test scores from neuroimaging data alone. It is increasingly common to track longitudinal changes over very short periods of time. In human neuroimaging, this interval has become as short as three months [1]. One may ask if there is even a "signal"

M.H. Coen and M. Hidayath Ansari—Contributed equally to this work.

© Springer International Publishing AG 2016
I. Rish et al. (Eds.): MLINI 2014, LNAI 9444, pp. 107–114, 2016.
DOI: 10.1007/978-3-319-45174-9_11

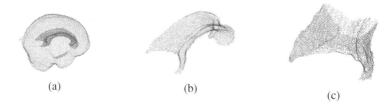

(a)   (b)

(c)

Fig. 1. (a) The blue outer mesh is a 3-D view of the surface of the human brain. The red inner mesh outlines the corpus callosum. (b) A view of the corpus callosum, a thick band of nerve fibers that connects the left and right hemispheres of the brain. (c) A view of the splenium of the corpus callosum, containing over 12,000 voxels. (Color figure online)

to find here. How do we know if there is anything meaningful to detect? This is exacerbated when the sampling time frame is much shorter than the onset time of observable phenomena we would like to predict.

Our approach begins with a "simple" classification problem. For longitudinal data, one instance of ground truth is the chronological order in which the datasets were collected. Thus, a natural question is: can we determine this order for a given individual? Solving this problem allows us to identify and rank the most temporally significant (*longitudinally*) and consistent (*cross-sectionally*) voxels. We hypothesize these voxels correlate with other temporally sensitive data, such as cognitive test scores. In confirming this hypothesis using the novel computational methods in Sect. 3 for the experiments in Sect. 4, we present the first evidence demonstrating that very small changes in white matter structure over a two year period can predict change in cognitive function in healthy adults.

2   Background and Data

Much previous research on Alzheimer's disease has focused on gray matter; white matter has historically been regarded as less relevant to cognition. In recent years, however, the role of white matter in the transfer of information has attracted vigorous interest [15] and is the subject of our paper. Data examined here come from the Wisconsin Registry for Alzheimer's Prevention [12]. Longitudinal imaging and cognitive testing data were available for 75 subjects, who were healthy, middle-aged (ranging from ages 45 to 70), and tested cognitively normal on neuropsychological assays. 78 % of subjects showed one or more risk factors for AD. All participants underwent comprehensive neuropsychological testing. The factor score of interest in this paper is the *Speed and Flexibility* factor [11,14].

We use measurements of white matter volume and structure from diffusion tensor magnetic resonance imaging (DT-MRI, or DTI). From these tensors, we derive the common scalar summary measure of Fractional Anisotropy (FA). Figure 1 illustrates the location and shapes of some regions of interest analyzed in this paper.

3   A Point Set Approach

In many classification tasks, data are often abstracted into a representation (e.g., a vector) that fails to retain their spatial information. Given the inherent spatial nature of voxel data, rather than serialize the voxels of a brain or region into one vector and lose their locations, we represent them as a point set $B = (V, W)$ where $V \subset \mathbb{R}^3, W \subset \mathbb{R}$ and every point v_i in V has a corresponding weight w_i in W. Weights w_i correspond to FA values here.

We compare DTI scans by defining a custom similarity between their respective voxel sets. This similarity is not a simple point-to-point similarity; rather it is between two point sets. One such measure of similarity is random Fourier features [9], in which a map $\tilde{\Phi}$ ("lifting" function) is applied to each data point in $\mathbb{R}^d$. This transforms it into an element of $\mathbb{R}^D$, a D-dimensional approximation of a reproducing kernel Hilbert space (RKHS). The mapping is randomized and similarity-preserving; a shift-invariant kernel in the original space is approximately equal to the inner product in the new space, where the approximation can be made as precise as possible by varying the dimensionality (D) of the lifted space. For the kernel $K(\boldsymbol{x}, \boldsymbol{y}) = e^{-\frac{\|\boldsymbol{x}-\boldsymbol{y}\|^2}{2}}$, the approximate lifting map $\hat{\Phi}_D : \mathbb{R}^d \to \mathbb{R}^D$ is defined as follows:

$$\hat{\Phi}(\boldsymbol{x}) = [cos(\omega_1 \boldsymbol{x}), \ldots, cos(\omega_{\frac{D}{2}} \boldsymbol{x}), sin(\omega_1 \boldsymbol{x}), \ldots, sin(\omega_{\frac{D}{2}} \boldsymbol{x})] \quad \text{for } \boldsymbol{x} \in \mathbb{R}^d$$

where elements of ω_i's are random and normally distributed, and $\langle \hat{\Phi}(\boldsymbol{x}), \hat{\Phi}(\boldsymbol{y}) \rangle \simeq K(\boldsymbol{x}, \boldsymbol{y})$ for any $\boldsymbol{x}, \boldsymbol{y} \in \mathbb{R}^d$.

Raman *et al.* [10] applied this approximate lifting map in representing point sets as elements of an RKHS. The map is applied to each point in a point set, and the whole set is then represented as a single vector by summing the lifted representations of the constituent points. The summed vector is normalized to unit length to eliminate differences caused by differing set cardinalities. The similarity between two point sets X and Y is defined as the dot product between the vectors representing them. We extend this formulation of point set similarity to incorporate weights for each point, so that the final expression for similarity between two point sets $X = (V_X, W_X)$ and $Y = (V_Y, W_Y)$ becomes

$$\langle \frac{\hat{\Phi}(X)}{\|\hat{\Phi}(X)\|}, \frac{\hat{\Phi}(Y)}{\|\hat{\Phi}(Y)\|} \rangle, \text{ where } \hat{\Phi}(X) = \sum_{v_i \in V_X} w_i \hat{\Phi}(\boldsymbol{v_i}).$$

3.1   Identifying Subsets of Informative Voxels

Given the large number of voxels in each scan, we combined longitudinal and cross-sectional data to identify those voxels that had comparatively *large, consistent*, and *similar* values in all difference images corresponding to a given class. Our hypothesis is that the voxels that change similarly in all subjects (cross-sectionally) across time (longitudinally) are the ones most sensitive to temporal ordering. Towards this, we define a "Q-value" for each voxel as follows:

$$Q(v_i) = \frac{\text{mean}(\text{FA}_i^1 - \text{FA}_i^2)}{\text{var}(\text{FA}_i^1 - \text{FA}_i^2)} \qquad (1)$$

where FA_i^1 is the FA value at voxel i at time 1, FA_i^2 the value at time 2, and mean and variance are computed cross-sectionally over the subject population.

We also define an additional quantity called CONSISTENCY (or CONS) for a voxel as follows, measuring the (higher) percentage of subjects who show the same sign change in that voxel from time 1 to time 2. We then set thresholds on Q and CONS and extract those voxels with values above these thresholds.

4   Experiments and Analysis

We present two experiments that demonstrate application of our framework to detecting minute, short-term changes in WM structure and relating them to changes in cognitive test scores.

4.1   Before Vs. After

The task in this experiment is to determine the temporal ordering in pairs of scans for an individual. We do this by exploiting data from voxels that undergo changes that are consistent and similar *across* subjects. This problem is challenging for several reasons: (1) The time period between scans is extremely short (1.5–2 years) and the subtle changes in the scans are believed to be largely unrelated to cognition; (2) All subjects are healthy and middle-aged and do not exhibit any pathology; (3) Domain experts in neuroscience and radiology are unable to solve this problem for healthy patients better than chance.

Fig. 2. Portions of the splenium of the corpus with high CONS value. Red voxels indicate a consistent increase in FA value across subjects, while blue represents decrease. (Color figure online)

For each of the 75 subjects, we subtract the latter image from the earlier one ("positive" difference images) and then reverse the order of subtraction ("negative" difference images). When given two new images from a single subject with no ordering information, we perform the subtraction in an arbitrary manner and compute which set of difference images this new difference image is more "similar" to, using the kernel in Sect. 3.

Since there are an equal number of positive and negative difference images, the baseline accuracy for this experiment is 50 %. We applied two simple mean-of-ROI (region of interest) classification methods for comparisons with our technique. The classification rule "the scan with the higher mean is the earlier image" achieves an accuracy rate of 57 % on the splenium of the corpus callosum - little better than random chance. This is because not all voxels show a decrease in FA value over time; in fact some voxels show an increase. A slightly more complex rule that weights FA changes by the signs of Q for each voxel yields an accuracy of 82 % for the same region, requiring all 12,729 voxels in the splenium.

Table 1. Classification results for predicting the before image from the later image using four different WM regions. The number of voxels reported is the mean across the different folds in each experiment.

Region	Number of voxels (CONS > 70 %)	Accuracy
Corpus callosum (whole)	3429 voxels	96 %
Corpus callosum (splenium)	463 voxels	97.3 %
Corpus callosum (genu)	364 voxels	90.7 %
Cingulum bundle (R & L)	776 voxels	97.3 %

Classification and Accuracy. We trained a support vector machine (SVM) with the kernel derived from Random Fourier Features (Sect. 3) to classify "positive" and "negative" difference images. Accuracy was determined with 10-fold cross validation, insuring no cross-leaking between test and training sets. The 10-fold cross-validation accuracy in predicting "before" scans from "after" scans (i.e. "positive" difference images from "negative" difference images) is shown for different WM regions in Table 1. As the table shows, approximately 400 well-chosen voxels are sufficient to achieve a classification accuracy of 96 %.

The identified voxels of interest are visualized in Fig. 2. Voxels can be distinguished based on whether they show an upward trend in FA value or a downward trend. The figure shows that voxels tend to be spatially proximal to other voxels of the same type. We note that this naturally-occurring "clustering" of nearby voxels with similar trends is readily apparent even when no smoothing is applied to the data. Further study of these regions and the trends within them (FA changes, demyelination, and relations to cognitive impairment) will be useful in understanding patterns of age-related change in FA and are discussed in Sect. 5.

4.2 Regression

We would like to model changes in subjects' neuropsychological test scores using FA differences observed over time. Even employing the Q score defined above to prune the space of voxels, the models remains extremely wide. Fitting multivariate linear models in this case cannot be done without constraints. While there are many familiar approaches limiting model exploration (e.g., lasso) and ways to validate them (e.g., residual distributions), evaluating the assessments is difficult with a limited number of samples. The data themselves are difficult to work with, as none of the differences between earlier and later test scores is statistically significant according to paired t-tests adjusted for inequality of variances. Scatterplots of earlier vs. later test scores fit lines of slope 1 with relatively high R^2. In these cases, even null models perform well.

To better manage the need for constrained variable selection with wide data, we used the coordinate descent approach for lasso and ridge [4]. We modified our approach to perform logistic regression on the *signs* of the test score changes,

Table 2. Classification results for predicting Speed and Flexibility from voxels

Method	Parameters	Accuracy (voxels)	Accuracy (clusters)
Lasso logistic regression [4]		70 % ($\lambda = .011$)	75 % ($\lambda = .013$)
SVM, Lifted kernel	$D = 500, C = 1$	58 %	55.7 %
SVM, Gaussian kernel	$\sigma = 1, C = 1$	57 %	58.5 %
Baseline random guessing		54 %	54 %

viewed as binomial distributions. Doing so normalizes the error penalty and allows us to pose a well-defined problem: do changes in neuroimaging data predict whether a subject's score for a neuropsychological test has increased or decreased? While one may suppose that cognitive abilities deteriorate monotonically with age, evidence does not bear this out, as discussed in Sect. 5. Our output variable is the sign of a small difference that appears to fluctuate around zero at random. However, lasso logistic regression via coordinate descent run 100 times with 10-fold cross validation achieved classification accuracy of 70 % with shrinkage parameter $\lambda = .011$ (within one standard error of the minimum). Results for this and other methods are shown in Table 2. No significant improvement was seen for other parameters on competing approaches.

These results are quite surprising. While achieving 70 % accuracy seems modest, consider that this prediction is made using voxel-based neuroimaging data selected because they were able to accurately answer our initial "Which image came first?" question. Within their own representation, the outcome data do not appear separable. But when viewing them from the neuroimaging perspective, we can classify them.

Clustering. In general, we prefer as few explanatory variables in a model as possible. Wide linear models always raise the specter of overfitting and are notoriously difficult to interpret, particularly when constructed with lasso. Following the spatial point set approach in Sect. 3, we cluster the voxels based on spatial proximity and their Q values. Linkage-based clustering connects voxels with their neighbors if their Q values are within ρ percent of each other. We typically take $\rho = 15$ and specify the maximum number of desired clusters as 30. Emerging from the clustering was the observation that spatially adjacent voxels are likely to have similar Q values. The regions corresponding to clustered voxels are shown in Fig. 3.

Because the clusters are internally consistent with respect to Q values, we used their mean FA values in a ridge logistic regression analysis to predict the sign of the change in the Speed and Flexibility score. Because the number of features (regions) is 30 here, we are no longer dealing with wide data, alleviating many of the concerns that they raise. The clusters are found to be better predictors than the voxels used in the previous model. Ridge logistic regression via coordinate descent run 100 times with 10-fold cross validation achieved a classification accuracy of 75 % for shrinking parameter $\lambda = 0.13$.

5   Discussion

This paper presents a new approach for longitudinal analysis of neuroimaging data. Our approach relies on the spatial nature of the data both for defining a new kernel and for clustering voxels based on their perceived quality or Q value. This kernel can be used to reliably classify scans based on small changes in their white matter structure – a task that eludes human experts. We then used the voxels that enabled this classification to predict changes in the significant cognitive factor of Speed and Flexibility. While a relationship between speed based cognitive tests and white matter microstructure has been *qualitatively* examined in cross-sectional studies, this is the first work to determine that change in FA over two years can predict change in cognitive function in healthy adults. The two experiments in this paper are on subject-level predictions, not group differences. As such, work on two-sample testing [5] is not relevant to this problem.

Fig. 3. A view of voxels clustered by Q values. Colors correspond to different clusters. (Color figure online)

Our results show that over time, portions of the splenium decrease in FA over a time period of approximately 2 years, which is expected in aging. More unexpected were white matter tracts that showed an increase in FA. (The red regions of Fig. 2.) The splenium of the corpus callosum carries fibers that connect the bilateral temporal, parietal and occipital lobes. It is possible that changes occurring over time include both loss of fibers and regenerative myelination [2]. The tight relationship found with the Speed and Flexibility factor score is not surprising because speed of neural conduction relies on intact myelin.

References

1. Alzheimer's disease neuroimaging initiative (ADNI). http://www.adni-info.org
2. Bowley, M.P., Cabral, H., Rosene, D.L., Peters, A.: Age changes in myelinated nerve fibers of the cingulate bundle and corpus callosum in the rhesus monkey. J. Comp. Neurol. **518**(15), 3046–3064 (2010)
3. Dyrba, M., et al.: Combining DTI and MRI for the automated detection of Alzheimer's disease using a large european multicenter dataset. In: Yap, P.-T., Liu, T., Shen, D., Westin, C.-F., Shen, L. (eds.) MBIA 2012. LNCS, vol. 7509, pp. 18–28. Springer, Heidelberg (2012)
4. Friedman, J., Hastie, T., Tibshirani, R.: Regularization paths for generalized linear models via coordinate descent. J. Stat. Soft. **33**(1), 1 (2010)
5. Gretton, A., Fukumizu, K., Sriperumbudur, B.K., et al.: A fast, consistent kernel two-sample test. In: Advances in Neural Information Processing Systems, pp. 673–681 (2009)
6. Grydeland, H., Westlye, L.T., Walhovd, K.B., Fjell, A.M.: Improved prediction of Alzheimer's disease with longitudinal white matter/gray matter contrast changes. Hum. Brain Mapp. (2012)

7. Le Bihan, D., Mangin, J.-F., Poupon, C., Clark, C.A., Pappata, S., Molko, N., Chabriat, H.: Diffusion tensor imaging: concepts and applications. J. Magn. Reson. Imaging **13**(4), 534–546 (2001)
8. Misra, C., Fan, Y., Davatzikos, C.: Baseline and longitudinal patterns of brain atrophy in mci patients, and their use in prediction of short-term conversion to AD: results from adni. Neuroimage **44**(4), 1415 (2009)
9. Rahimi, A., Recht, B.: Random features for large-scale kernel machines. Adv. Neural Inf. Process. Syst. **20**, 1177–1184 (2007)
10. Raman, P., Phillips, J.M., Venkatasubramanian, S.: Spatially-aware comparison and consensus for clusterings. In: Proceedings of SIAM International Conference on Data Mining (SDM) (2011)
11. Reitan, R.M., Wolfson, D.: The halstead-reitan neuropsychological test battery for adults: theoretical, methodological, and validational bases. In: Neuropsychological Assessment of Neuropsychiatric and Neuromedical Disorders, p. 1 (2009)
12. Sager, M.A., Hermann, B., La Rue, A.: Middle-aged children of persons with alzheimer's disease: APOE genotypes and cognitive function in the wisconsin registry for alzheimer's prevention. J. Geriatr. Psychiatry Neurol. **18**(4), 245–249 (2005)
13. Smith, S.M., Zhang, Y., Jenkinson, M., Chen, J., Matthews, P.M., Federico, A., De Stefano, N., et al.: Accurate, robust, and automated longitudinal and cross-sectional brain change analysis. Neuroimage **17**(1), 479–489 (2002)
14. Trenerry, M.R., Crosson, B., DeBoe, J., Leber, W.R.: Stroop Neuropsychological Screening Test Manual. Psychological Assessment Resources, Odessa (1989)
15. Ziegler, D.A., Piguet, O., Salat, D.H., Prince, K., Connally, E., Corkin, S.: Cognition in healthy aging is related to regional white matter integrity, but not cortical thickness. In: Neurobiology of Aging, vol. 31, pp. 1912–1926. Elsevier, November 2010

Hyperalignment of Multi-subject fMRI Data by Synchronized Projections

Raif M. Rustamov$^{(\boxtimes)}$ and Leonidas Guibas

Computer Science Department, Stanford University, Stanford, USA
rustamov@research.att.com

Abstract. Group analysis of fMRI data via multivariate pattern methods requires accurate alignments between neuronal activities of different subjects in order to attain competitive inter-subject classification rates. Hyperalignment, a recent technique pioneered by Haxby and collaborators, aligns the activations of different subjects by mapping them into a common abstract high-dimensional space. While hyperalignment is very successful in terms of classification performance, its "anatomy free" nature excludes the use of potentially helpful information inherent in anatomy. In this paper, we present a novel approach to hyperalignment that allows incorporating anatomical information in a non-trivial way. Experiments demonstrate the effectiveness of our approach over the original hyperalignment and several other natural alternatives.

1 Introduction

Apart from being a fundamental issue in cognitive neuroscience, "the problem of conceptual similarity across neural diversity" [3] has a direct practical manifestation when analyzing fMRI data. Namely, group analysis of fMRI data via multivariate pattern methods requires aligning activations of different subjects. While, pragmatically, the goal of alignment is to attain inter-subject classification (ISC) rates comparable to within subject classification (WSC) rates, ideally, such alignments should take into account both anatomical and functional features of the brain.

Existing spatial alignment approaches are based either purely on anatomical features [6,12], or on a combination of anatomical features with features extracted from fMRI data, such as activations directly [9] or connectivity derived from activations [4,5]. However, these approaches do not consistently yield ISC rates comparable to WSC rates [5]. On the opposite end of the spectrum is the recently introduced class of methods summed under the name "hyperalignment" [7,8,14]. Hyperalignment essentially finds linear combinations of voxel activations that agree across the subjects, yielding subject specific linear maps (matrices) that transform their activations into a common abstract high-dimensional space. While hyperalignment works well in practice achieving ISC rates on par with or even better than WSC rates, in the current form, it lacks a mechanism for incorporating anatomical information that potentially may lead to even better classification performance.

© Springer International Publishing AG 2016
I. Rish et al. (Eds.): MLINI 2014, LNAI 9444, pp. 115–121, 2016.
DOI: 10.1007/978-3-319-45174-9_12

The goal of this work is to introduce an approach to hyperalignment that allows the use of anatomical information. We start by computing pair-wise (hyper-) alignments between subjects by setting up an optimization problem containing terms involving both anatomical and functional features. Next, we need to aggregate these pair-wise alignments into an overall alignment of all subjects. To achieve this, inspired by the recent work on synchronization [10,11,13], we introduce the method of synchronized projections, which yields the final maps of activations into a common space shared between all subjects.

Our approach has a number of advantages. First, while our approach shares the same core idea with hyperalignment – mapping activations into a common space – yet our maps are heavily guided by anatomical information. Second, we do not make any restrictions on the choice of pair-wise alignments; the synchronized projections can be applied more generally to any set of pair-wise alignments that can be expressed as linear maps. Third, experimental results confirm the superiority of synchronized projections in terms of ISC rates over more straightforward approaches that align all subjects to a reference subject or to a floating subject that is iteratively refined.

The paper is organized as follows. We introduce our computation procedure for pair-wise alignments in Sect. 2.1. The main technical contribution, the method of synchronized projections, is described in Sect. 2.2. We present experimental evaluation of our approach on a multi-subject category perception data in Sect. 3.

2   Approach

The input to our algorithm is fMRI data elicited from n_{subj} subjects exposed to a common synchronous stimulus, such as viewing a number of images in the same order. The data for i-th subject is recorded in $n_{\text{TR}} \times n_{\text{vox}}$ matrix X^i, where each row corresponds to a time point, and each column to a voxel in the subject's brain. Note that each row-vector represents a spatially-varying fMRI activation at some time, and the rows in X^i are ordered consistently across all subjects. On the other hand, the columns – each containing the time course of a particular voxel – are *not* assumed to be in correspondence across the subjects. Since the activations of different subjects are not directly comparable, we cannot train a single multi-voxel pattern classifier that would work for all the subjects at once.

Our goal is to provide a way of computing features/projections of fMRI activations that are consistent across subjects. To this end, our algorithm computes projection matrices – one for each subject – which can be used to map activations of that subject into a common space shared between all subjects.

The algorithm proceeds in two steps. First, for each pair of subjects, we compute a linear map that transports the activation vectors of one subject to the reference frame of the other. In the second step, we compute the projection matrices by setting up an optimization problem which essentially requires the following: the projection of an activation should be roughly the same if one were to transport the activation to another subject and then project. This leads to a matrix eigenvalue problem for some symmetric positive semi-definite matrix. The details of our construction are provided in the remainder of this section.

2.1 Pair-Wise Alignment Maps

As in original hyperalignment, we will use linear maps to align fMRI responses of different subjects. Thus, the first step of our algorithm is to compute, for every pair of subjects $i, j = 1, ..., n_{\text{subj}}$, a linear map (matrix) C^{ij} that transforms fMRI activations of subject i to the reference frame of subject j, namely by achieving $X^i C^{ij} \approx X^j$. What this means is that while the voxel activations in two different subjects may not be directly comparable, yet one can make linear combinations from activations of some voxels in the i-th subject's brain that will be compatible with the activation of a given voxel in the j-th subject's brain. Thus, the matrix entry C_{pq}^{ij} captures the coefficient with which voxel q of subject i appears in the linear combination for voxel p of subject j.

These alignment matrices are learned from the training data by posing an optimization problem of the following form: $C^{ij} = \arg\min_C \|X^i C - X^j\|_F$, where $\|\cdot\|_F$ is the Frobenius norm. Since the amount of training data is limited, this optimization problem is overly under-determined and needs some kind of regularization. For example, the original hyperalignment [7] requires the matrices C^{ij} to be orthogonal.

Here we propose a different regularization that incorporates the anatomical information. Remember that the brains can be anatomically aligned using a number of approaches; here we will use the Talairach alignment [12]. As a result of such alignment, all of the brain images are placed into a common 3D space, and one can compute the Euclidean distance D_{pq}^{ij} between voxel q of subject i and voxel p of subject j. We now seek the pairwise alignment matrix via the following optimization problem:

$$C^{ij} = \arg\min_C \|X^i C - X^j\|_F^2 + \mu \sum_{p,q} (D_{pq}^{ij} C_{pq})^2. \tag{1}$$

The proposed regularization term has an important advantage over the orthogonality requirement of original hyperalignment. Orthogonality requirement makes it possible for spatially distant voxels to take part in the linear combination for a given voxel, rendering hyperalignment "anatomy free". Our regularizer, on the other hand, penalizes spatially distant voxels, effectively imposing the prior that the anatomical alignment is not too far from truth.

2.2 Synchronized Projections

The second step of our algorithm uses the pairwise alignment maps in order to construct projection matrices into a d-dimensional common space shared between all subjects. To this end, for each subject i we construct an $n_{\text{vox}} \times d$ matrix P^i, such that the projected activations $X^i P^i$ are consistent between subjects and can be used to train a single multi-voxel pattern classifier that would work for all subjects at once.

The main idea is as follows: if we have an activation row-vector v of subject i, then it can be transported to subject j by computing vC^{ij}; since the activation before and after transport represents the same stimulus, then their projections (using the respective subject's projection matrix) should be roughly the same: $vC^{ij}P^j \approx vP^i$. Since this should hold for all activation vectors and all pairs of subjects, we can setup an optimization problem that minimizes the discrepancies between projections. One way to formalize this is to seek the projection matrices as minimizers of $\sum_{i,j} \|C^{ij}P^i - P^j\|_F^2$, subject to normalization constraints to avoid trivial solutions.

To put the problem into a more familiar form, let us denote by $\mathbb{P}$ the $n_{\text{subj}}n_{\text{vox}} \times d$ matrix obtained by stacking together all of the matrices $P^i, i = 1, ..., n_{\text{subj}}$. We can rewrite the optimization objective as follows:

$$\sum_{i,j} \|C^{ij}P^i - P^j\|_F^2 = \mathbb{P}^{\top}\mathbb{L}\mathbb{P}, \tag{2}$$

where $\mathbb{L}$ is $n_{\text{subj}}n_{\text{vox}} \times n_{\text{subj}}n_{\text{vox}}$ matrix. This matrix consists of $n_{\text{subj}} \times n_{\text{subj}}$ blocks L^{ij} of size $n_{\text{vox}} \times n_{\text{vox}}$. Namely, letting I be the $n_{\text{vox}} \times n_{\text{vox}}$ identity matrix, we have

$$L^{ij} = \begin{cases} -(C^{ij} + C^{ji\top}) & , i \neq j \\ (n_{\text{subj}} - 1)I + \sum_{k,k\neq i} C^{ki\top}C^{ki} & , i = j \end{cases}$$

As can be seen directly from Eq. (2), $\mathbb{L}$ is a symmetric positive semi-definite matrix. This is a generalized notion of graph Laplacian to the setting where edges are decorated by pair-wise mappings [10,11,13].

Since our goal is to minimize $\mathbb{P}^{\top}\mathbb{L}\mathbb{P}$, we need to impose some constraints on $\mathbb{P}$ in order to avoid trivial solutions. We require the columns of $\mathbb{P}$ to be orthonormal because this leads to an eigenvalue problem. Namely, it is easy to see that then the columns of optimal $\mathbb{P}$ are simply the eigenvectors corresponding to the smallest d eigenvalues of $\mathbb{L}$, and that the optimal objective value is the sum of the smallest d eigenvalues of $\mathbb{L}$.

It follows that when the dimension d of the common space is increased from one value to another, all the previous projected coordinates are kept intact and new coordinates are added. In a sense, the projected coordinates are naturally ordered by their corresponding eigenvalues – the smaller the eigenvalue, the stronger is the inter-subject commonality (i.e. the smaller is its contribution to the discrepancy as measured by our objective) captured by the corresponding projected coordinate. Therefore, to obtain a low-dimensional common representation space, we do not need to start with a high-dimensional space and then select the principal component directions as in original hyperalignment [7]. Our eigenvalue based ordering of coordinates provides a more principled criterion than the maximum variance directions criterion of the PCA, because large variance could in fact be due to the absence of commonality along a direction.

3   Results

Our goal is to show the benefit of synchronization in comparison to two other natural approaches, and also to compare it with anatomical alignment and the

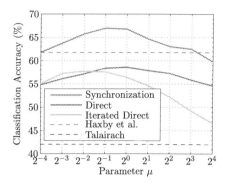
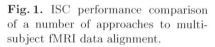

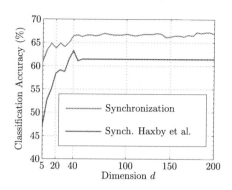

Fig. 1. ISC performance comparison of a number of approaches to multi-subject fMRI data alignment.

Fig. 2. Dependence of ISC performance on the dimension d of the common space.

original hyperalignment of [7]. Our experiments are based on category perception (faces and objects) data from [7] that is distributed together with hyperalignment module [1] of PyMVPA package. This dataset is challenging as evidenced, for example, by the inability of a previously introduced generalization of hyperalignment [14] to improve over the original hyperalignment [1]. This can be attributed in part to the small size of the dataset, which limits the number of training samples.

Our evaluation protocol follows directly the one described in [7]: first, for all subjects, all runs except one are used for voxel selection, pair-wise map computation, and determining common space; second, a linear multi-class SVM classifier [2] is trained on these runs of all subjects except one subject; the classifier is tested on the held-out run of the held-out subject. The obtained classifier accuracy is a measure of inter-subject classification success; this accuracy, averaged over held-out subjects and held-out runs, constitutes our performance metric.

Figure 1 shows the performance comparison of five different approaches to alignment of multi-subject fMRI data. The horizontal axis on this graph is the value of the regularization weight μ appearing in the optimization problem for computing pair-wise maps, Eq. (1). Of course, the performances of Talairach alignment (taken directly from the hyperalignment module website [1]) and original hyperalignment [7] (re-implemented in MATLAB for consistency; results are in agreement with hyperalignment module [1]) are independent of the parameter μ. In accordance with [7], voxel selection is done by retaining a fixed number ($n_{\text{vox}} = 200$) of voxels with highest F-scores. For all methods in this figure except synchronization, the dimension of common space is tied (equal) to the number of voxels; for fair comparison, we set $d = n_{\text{vox}}$ for synchronization as well.

The performance of our synchronization approach is also compared to two other natural approaches, labeled "direct" and "iterated direct" in the graph. The direct approach picks one of the subjects, say r, as a reference, and then uses the pair-wise maps C^{ir} to map the activations of all of the subjects to the frame of this reference subject; these mapped activations are used as features in machine learning step. The iterated direct approach starts out in exactly the same manner, except that mapping process is repeated. Namely, after the first mapping is complete, for each TR, the average of mapped activations are computed, and the new pair-wise maps (from all subjects to the reference subject) are computed to match these averaged activations on reference subject. This iterative process is similar to the original hyperalignment technique of Haxby et al. [7], except that the pair-wise maps are computed using Eq. (1). The performances of direct and iterated direct approaches are averaged over all the reference subject choices.

As indicated by Fig. 1, the synchronization approach consistently improves over the natural alternatives – the direct and iterated direct approaches – using the same type of pair-wise maps. In addition, for some parameter settings, synchronization provides a non-negligible improvement over the original hyperalignment approach of Haxby et al. [7].

Next, we fix the parameter $\mu = 1$, and study the dependence of ISC performance on the dimension d of the common space. Figure 2 shows that even for the dimensionality as low as 10, our approach yields performance competitive with Haxby et al. hyperalignment. This is in agreement with the finding in [7] that keeping a limited number of principal components of the common space is sufficient for obtaining improved ISC rates. However, here we do not need to apply principal component analysis, because the coordinates of common space obtained via our algorithm are already ordered by the degree of inter-subject commonality; see discussion at the end of Sect. 2.2.

Finally, we investigate what happens if one were to change the type of pair-wise alignments used in synchronization. Following the idea of original hyperalignment [7], we require that the pair-wise alignment matrices are orthogonal. More precisely, in optimization problem of Eq. (1) we drop the anatomy based regularizer, and instead require that C^{ij} is orthogonal, which reduces the problem to Procrustes analysis as in [7]. The curve in Fig. 2 labeled "Synch. Haxby et al." shows the performance of synchronization applied to these new pair-wise maps. It can be seen that the performance is equivalent to the original hyperalignment starting at around dimension $d = 35$, which is in good agreement with the dimension of reduced common space identified in [7] via PCA.

4   Conclusion

We have introduced an approach allowing to inject anatomical information into hyperalignment. Experiments demonstrated the effectiveness of our approach over the original hyperalignment and several other natural alternatives.

Acknowledgments. The authors acknowledge the support of NSF grants FODAVA 808515 and DMS 1228304, AFOSR grant FA9550-12-1-0372, ONR grant N00014-13-1-0341, a Google research award, and the Max Plack Center for Visual Computing and Communications.

References

1. PyMVPA User Manual: Hyperalignment for Between-Subject Analysis. http://dev.pymvpa.org/examples/hyperalignment.html
2. Chang, C.C., Lin, C.J.: LIBSVM: a library for support vector machines. ACM Trans. Intell. Syst. Technol. **2**, 27:1–27:27 (2011). http://www.csie.ntu.edu.tw/ cjlin/libsvm
3. Churchland, P.M.: Conceptual similarity across sensory and neural diversity: the Fodor/Lepore challenge answered. J. Philos. **95**(1), 5–32 (1998)
4. Conroy, B.R., Singer, B., Haxby, J.V., Ramadge, P.J.: fMRI-based inter-subject cortical alignment using functional connectivity. In: NIPS, pp. 378–386 (2009)
5. Conroy, B.R., Singer, B.D., Guntupalli, J.S., Ramadge, P.J., Haxby, J.V.: Inter-subject alignment of human cortical anatomy using functional connectivity. NeuroImage **81**, 400–411 (2013)
6. Fischl, B., Sereno, M.I., Tootell, R.B., Dale, A.M.: High-resolution intersubject averaging and a coordinate system for the cortical surface. Hum. Brain Mapp. **8**(4), 272–284 (1999)
7. Haxby, J.V., Guntupalli, J.S., Connolly, A.C., Halchenko, Y.O., Conroy, B.R., Gobbini, M.I., Hanke, M., Ramadge, P.J.: A common, high-dimensional model of the representational space in human ventral temporal cortex. Neuron **72**(2), 404–416 (2011)
8. Lorbert, A., Ramadge, P.J.: Kernel hyperalignment. In: NIPS, pp. 1799–1807 (2012)
9. Sabuncu, M.R., Singer, B.D., Conroy, B., Bryan, R.E., Ramadge, P.J., Haxby, J.V.: Function-based intersubject alignment of human cortical anatomy. Cereb. Cortex **20**(1), 130–140 (2010)
10. Singer, A.: Angular synchronization by eigenvectors and semidefinite programming. Appl. Comput. Harmon. Anal. **30**(1), 20–36 (2011)
11. Singer, A., Wu, H.T.: Vector diffusion maps and the connection Laplacian. Comm. Pure Appl. Math. **65**(8), 1067–1144 (2012)
12. Talairach, J., Tournoux, P.: Co-planar Stereotaxic Atlas of the Human Brain: 3-D Proportional System: An Approach to Cerebral Imaging. Thieme, Stuttgart (1988)
13. Wang, F., Huang, Q., Guibas, L.: Image co-segmentation via consistent functional maps. In: Proceedings of the International Conference on Computer Vision (ICCV) (2013)
14. Xu, H., Lorbert, A., Ramadge, P., Guntupalli, J., Haxby, J.: Regularized hyperalignment of multi-set fmri data. In: IEEE Statistical Signal Processing Workshop, pp. 229–232 (2012)

An Oblique Approach to Prediction of Conversion to Alzheimer's Disease with Multikernel Gaussian Processes

Jonathan Young[1]([✉]), Marc Modat[1], Manuel J. Cardoso[1], John Ashburner[2], and Sebastien Ourselin[1,3]

[1] Centre for Medical Image Computing, University College London, London, UK
jonathan.young@ucl.ac.uk
[2] Wellcome Trust Centre for Neuroimaging, Institute of Neurology,
University College London, London, UK
[3] Dementia Research Centre, Institute of Neurology,
University College London, London, UK

Abstract. Machine learning approaches have had some success in predicting conversion to Alzheimer's Disease (AD) in subjects with mild cognitive impairment (MCI), a less serious condition that nonetheless is a risk factor for AD. Predicting conversion is clinically important as because novel drugs currently being developed require administration early in the disease process to be effective. Traditionally training data are labelled with discrete disease states; which may explain the limited accuracies obtained as labels are noisy due to the difficulty in providing a definitive diagnosis of Alzheimer's without post-mortem confirmation, and ignore the existence of a continuous spectrum of disease severity. Here, we dispense with discrete training labels and instead predict the loss of brain volume over one year, a quantity that can be repeatably and objectively measured with the boundary shift integral and is strongly correlated with conversion. The method combines MRI and PET image data and cerebrospinal fluid biomarker levels in an Bayesian multi-kernel learning framework. The resulting predicted atrophy separates converting and non-converting MCI subjects with 74.6 % accuracy, which compares well to state of the art methods despite a small training set size.

Keywords: Gaussian processes · Regression · Atrophy · BSI · Multi-kernel learning · MRI · PET Alzheimer's disease · Mild cognitive impairment

1 Introduction

In the study of AD, in recent years an increasing emphasis has been placed on the importance of early diagnosis. This is because while currently available treatments are only able to mitigate the downstream effects of the disease process, pending ones are focused on actually disrupting the disease process itself, by interfering with the amyloid cascade that is thought to be one of the underlying causes of AD [1]. To be effective, such treatments would have to begin

© Springer International Publishing AG 2016
I. Rish et al. (Eds.): MLINI 2014, LNAI 9444, pp. 122–128, 2016.
DOI: 10.1007/978-3-319-45174-9_13

before the patient is showing the full symptoms of AD. This involves studying patients who have mild cognitive impairment (MCI). Clinically, MCI is defined as having isolated memory deficits that are not severe enough to affect normal living [2]. MCI patients convert to AD at an annual rate of 10–15% per year [3], although some develop other diseases or remain stable. As stable and converting MCI (MCI-s and MCI-c) patients by definition have similar symptoms, standard cognitive tests used to diagnose AD are by themselves of little help for this problem; instead, imaging and other biomarkers can be used with machine learning methods to detect subtle differences between the groups. A classifier can be trained on labeled examples of MCI-s and MCI-c images, or alternatively on examples of AD patients and healthy controls (HC), under the assumption that MCI-s subjects are more HC like and MCI-c subjects are more AD like. Most such studies use magnetic resonance imaging (MRI), from which a variety of features can be extracted. However the results can be improved by combining MRI features with imaging data measuring metabolic activity using fluorodexoxyglucose positron emission tomography (FDG-PET) and biomarkers measured in a sample of cerebrospinal fluid (CSF) or genetic information in a multi-kernel framework [4,5].

A limiting factor in the accuracy these studies may be mislabeling of training subjects. The gold standard for diagnosis of AD is autopsy, but most studies use subjects whose diagnosis has been determined by standard clinical testing, which has been shown to have an error rate of at least 10 % [6]. This is an issue than has not been widely adressed; Aksu et al. [7] point out that training labels for MCI-s and MCI-c are uncertain and go on to generate their own MCI training labels by following the classification of MCI subjects by an HC versus AD classifier across multiple timepoint. However even this neglects the uncertainty in the HC and AD labels this scheme ultimately depends on.

Our proposed method follows [8] in abandoning discrete disease state labels for training altogether. We also perform a regression to predict a continuous proxy for disease status, but instead of age we use atrophy over a period of one year as measured by the boundary shift integral (BSI) [9]. This then provides a predicted atrophy rate for each test subject. We use Gaussian process (GP) regression [10], with a multiple kernel framework to optimally combine MRI, FDG-PET and CSF data. This results in a measure that can predict MCI conversion within 3 years with a balanced accuracy of 74.6 %, as good as state of the art techniques having a much larger training set, including our own previous work using multikernel GPs for classification [5].

2   Materials and Methods

2.1   Image and Biomarker Data

All data were obtained from the Alzheimer's Disease Neuroimaging Initiative (ADNI) database[1]. The MRI images were T1 weighted structural scans from

---

[1] http://adni.loni.ucla.edu/.

1.5T scanners taken at baseline and 12 month follow-up. All were subjected to quality control and automatically corrected for spatial distortion caused by gradient nonlinearity and B1 field inhomogeneity.

FDG-PET images were acquired according to the ADNI protocol: acquired 30–60 min post-injection, averaged, spatially aligned, interpolated to a standard voxel size, intensity normalized, and smoothed to a common resolution of 8-mm full width at half maximum.

CSF samples were obtained from subjects by a lumbar puncture around the time of their baseline scan. Levels of the proteins amyloid-β_{42} (aβ_{42}), tau, and phosphorylated tau were measured and recorded.

The original ADNI project collected baseline structural MRI scans for all subjects. However FDG-PET scanning and collection of CSF data were only done on subsets of these subjects. Furthermore, calculation of BSI requires a 12-month follow-up structural MRI, which were also missing for some subjects. As our method requires FDG-PET and CSF and a 12-month BSI as well as structural MRI data, only 129 subjects could be included in the study. The details of these are shown in Table 1. Subjects were classified as HC, AD or MCI by neuropsychological and clinical testing at the time of the baseline scan, with MCI conversion status decided by whether subjects were subsequently diagnosed as AD at any stage during the 36 month follow-up period.

Table 1. Subject groups and demographics

Disease status	Number	Female	Mean age (sd)
HC	28	19	74.1 (4.5)
MCI-s	38	22	75.3 (7.3)
MCI-c	29	18	75.1 (7.4)
AD	34	23	75.1 (6.8)

2.2   Image Processing

Probabilistic grey matter (GM) maps were produced from the native space baseline scans using the NiftySeg tool [11]. The native space images were also anatomically parcelated into 83 regions with a novel label fusion algorithm [12] in a multi-atlas label propagation scheme. The resulting parcelations were used to mask out the brainstem and cerebellum from the native space GM segmentations.

Also, a custom template was produced, performing all registrations using the NiftyReg toolkit [13]. The native GM space segmentations were then warped into the groupwise space. Finally, the groupwise space, masked GM segmentations were modulated by the Jacobian determinants of this final deformation. This step ensures the total mass of tissue remains constant. Hence the MRI features used were voxel level GM probabilities.

The native space anatomical parcelations were also rigidly transferred to the space of the FDG-PET images for the corresponding subjects. The parcelation was used to normalise each FDG-PET image by its mean cerebellar activity, and then to calculate the mean activity within each anatomical region, generating a set of 83 features for each FDG-PET image.

2.3 Boundary Shift Integral

The BSI is a method for robustly assessing volume loss of whole brains or brain regions from structural MRI. It calculates a change in volume by integrating across the longitudinal change in position of the boundary between CSF and GM surrounding the region of interest. Preprocessing is needed to extract the region of interest (which in our case is the whole brain) from each image, linearly align the baseline and follow-up images, and correct for intensity inhomogeneity between scans. We use the latest version of BSI [9] which uses a symmetric registration scheme to minimise bias and maximise desirable qualities for an atrophy measurement such as inverse consistency and transitivity between multiple time-points.

We normalise the resulting volume changes by the baseline brain volumes and by the exact interval between baseline and follow-up scans, and multiply by 100. This produces a normalised brain atrophy rate (BAR) in percentage of original brain volume per year for each subject. These are then used as targets in the following regression analysis.

2.4 Gaussian Process Regression

Gaussian processes (GPs) provide a Bayesian, kernelised framework for solving both regression and classification problems. As an in depth explanation of GPs is beyond the scope of this paper, we refer the reader to [10] for a theoretical treatment and our previous work [5] for an application of multi-kernel GPs to predicting conversion to AD.

Briefly, however, a GP is a multivariate Gaussian forming the prior on the value of a latent function, on top of which is put a Gaussian noise model. The covariance of the prior is a function of the covariance between instances of training data X, and a set of hyperparameters θ that control the overall form of the prior and the noise variance. During the training phase, the hyperparameters are learned from the training data X and targets y by type-II maximum likelihood. Once the hyperparameters have been set, predictions on unseen data are made by integrating across this prior, which can be calculated analytically for the regression case.

2.5 Gaussian Processes as Multimodal Kernel Methods

GP regression is based on a covariance kernel K, a symmetric positive definite matrix where entry K_{ij} is given by a covariance kernel function k of the feature vectors for the ith and jth subject $\boldsymbol{x}_i$ and $\boldsymbol{x}_j$ and a hyperparameter or hyperparameters θ. We use a linear kernel function, which is simply the scalar product of $\boldsymbol{x}_i$ and $\boldsymbol{x}_j$. As GPs belongs to the family of kernel methods, a positive sum of valid kernels is a valid kernel, and a valid kernel multiplied by a positive scalar is also a valid kernel. This implies that to do multimodal classification, we can define our kernel function as the weighted sum of a number of subkernels, each of which has been calculated from a the feature vectors representing a particular type of data or modality for each subject. Each subkernel has a scaling hyperparameter α representing the modality's weight in the overall kernel. A bias term β is also included in the sum. So in the case of multimodal classification using information derived from the MRI, PET and CSF data for each subject the overall kernel is

$$K_{ij} = \alpha_{\mathrm{MRI}}(\boldsymbol{x}_{i.\mathrm{MRI}}.\boldsymbol{x}_{j.\mathrm{MRI}}) + \alpha_{\mathrm{PET}}(\boldsymbol{x}_{i.\mathrm{PET}}.\boldsymbol{x}_{j.\mathrm{PET}}) + \alpha_{\mathrm{CSF}}(\boldsymbol{x}_{i.\mathrm{CSF}}.\boldsymbol{x}_{j.\mathrm{CSF}}) + \beta \quad (1)$$

giving a total of four covariance hyperparameters to set.

3   Results

To generate predicted BARs for all 129 subjects, we perform a leave-one-out cross validation (LOOCV) across the entire set. The correlation coefficient between predicted and measured BARs for the subjects is 0.38 (p < 0.0001) and the root mean squared error is 0.61. However our primary focus is not on the predicted brain atrophy rates themselves, but on whether they can be used to predict conversion in MCI subjects. Figures 1 and 2 show the spread of both measured and predicted BAR values for all four disease groups (HC, MCI-s, MCI-c, AD).

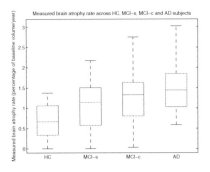
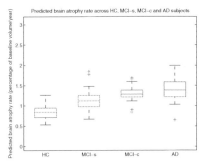

Fig. 1. Measured BAR across groups   **Fig. 2.** Predicted BAR across groups

As shown in Figs. 1 and 2, while the mean predicted BARs for each group are similar to the corresponding means for measured BARs, each clinical group occupies a much tighter cluster of values, even allowing for a few outliers (marked as

Table 2. Accuracy of discrimination between MCI-s and MCI-c with predicted BAR

Modalities	Accuracy (%)	AUC
MRI	59.7	0.595
PET	73.1	0.777
CSF	52.2	0.545
MRI, PET, CSF	74.6	0.725

Table 3. Accuracy of discrimination between MCI-s and MCI-c with training on binary diagnostic class labels

Training	Accuracy (%)	AUC
MCI (CV)	40.3	0.401
HC, MCI, AD (CV)	52.2	0.569
HC, AD	55.2	0.661

a +). This results in reduced overlap between the clinical groups, which is especially noticeable between the MCI-s and MCI-c groups. The resulting accuracy is 74.6 %, which is similar to the best previously reported results. The balanced accuracy and area under the ROC curve (AUC) are shown in Table 2. This also shows results for single modalities, demonstrating the benefit of combining sources of data with multikernel learning.

We also compare our method to performing direct binary classification on the conversion status again using GPs. This can be done by training on the MCI subjects only in an LOOCV loop, by training on all subjects, again with an LOOCV loop and grouping HC subjects with MCI-s and MCI-c subjects with AD, and finally by training on the HC and AD subjects, and testing on the MCI subjects. The results are given in Table 3.

4   Discussion

These results show a clear advantage for our method of training on a well-characterised proxy for MCI conversion, rather than the diagnostic status itself. Training on BAR enables us to reach accuracies of up to 74.6 %, whereas training on diagnostic labels struggles to perform better than chance. It therefore appears that the use of BAR bypasses the problems caused by binary diagnostic labels. Data is made better use of as subjects can be used for training regardless of diagnostic label, and as parameters are learned automatically there is no need to set subjects aside for tuning. We also show an advantage for multimodal regression. Although direct comparisons between methods are difficult [5], the resulting accuracy in forecasting MCI conversion is among the best yet achieved. The main drawback of our the proposed method is that all three types of data are all required for the best results (although FDG-PET alone does almost as well) which limits the number of subjects that can be included. However we intend to further evaluate this method as much greater numbers of subjects with all modalities become available in ADNI 2. Finally, while 12-month follow-up scans are also required to calculate BSI values for training data, it should be emphasised they are *not* needed for testing data.

128   J. Young et al.

Acknowldegments. We would like to thank Dr Kelvin Leung of the Dementia Research Centre, University College London for his assistance and provision of BSI data.

References

1. Robert, R., Wark, K.L.: Engineered antibody approaches for Alzheimer's disease immunotherapy. Arch. Biochem. Biophys. **526**(2), 132–138 (2012)
2. Petersen, R.C., Smith, G.E., Waring, S.C., Ivnik, R.J., Tangalos, E.G., Kokmen, E.: Mild cognitive impairment: clinical characterization and outcome. Arch. Neurol. **56**(3), 303–308 (1999)
3. Braak, H., Braak, E.: Staging of Alzheimer's disease-related neurofibrillary changes. Neurobiol. Aging **16**(3), 271–278 (1995)
4. Zhang, D., Wang, Y., Zhou, L., Yuan, H., Shen, D.: Multimodal classification of Alzheimer's disease and mild cognitive impairment. NeuroImage **55**(3), 856–867 (2011)
5. Young, J., Modat, M., Cardoso, M.J., Mendelson, A., Cash, D., Ourselin, S.: Accurate multimodal probabilistic prediction of conversion to Alzheimer's disease in patients with mild cognitive impairment. NeuroImage: Clin. **2**, 735–745 (2013)
6. Beach, T.G., Monsell, S.E., Phillips, L.E., Kukull, W.: Accuracy of the clinical diagnosis of Alzheimer disease at national institute on aging Alzheimer disease centers, 2005–2010. J. Neuropathol. Exp. Neurol. **71**(4), 266–273 (2012)
7. Aksu, Y., Miller, D.J., Kesidis, G., Bigler, D.C., Yang, Q.X.: An MRI-Derived definition of MCI-to-AD conversion for long-term, automatic prognosis of MCI patients. PLoS ONE **6**(10), e25074 (2011)
8. Gaser, C., Franke, K., Klöppel, S., Koutsouleris, N., Sauer, H.: Alzheimer's disease neuroimaging initiative: BrainAGE in mild cognitive impaired patients: predicting the conversion to Alzheimer disease. PLoS ONE **8**(6), e67346 (2013)
9. Leung, K.K., Ridgway, G.R., Ourselin, S., Fox, N.C.: Consistent multi-time-point brain atrophy estimation from the boundary shift integral. NeuroImage **59**(4), 3995–4005 (2012)
10. Rasmussen, C.E., Williams, C.K.I.: Gaussian Processes for Machine Learning. MIT Press, Cambridge (2006)
11. Cardoso, M.J., Clarkson, M.J., Ridgway, G.R., Modat, M., Fox, N.C., Ourselin, S.: LoAd: a locally adaptive cortical segmentation algorithm. NeuroImage **56**, 1386–1397 (2011)
12. Cardoso, M., Modat, M., Ourselin, S., Keihaninejad, S., Cash, D.: Multi-STEPS: multi-label similarity and truth estimation for propagated segmentations. In: 2012 IEEE Workshop on Mathematical Methods in Biomedical Image Analysis (MMBIA), pp. 153–158, January 2012
13. Modat, M., Ridgway, G.R., Taylor, Z.A., Lehmann, M., Barnes, J., Hawkes, D.J., Fox, N.C., Ourselin, S.: Fast free-form deformation using graphics processing units. Comput. Methods Programs Biomed. **98**(3), 278–284 (2010)

Author Index

Printed in the United States
By Bookmasters